Mohammad Ebrahim Azari Najaf Abad

# Optimization and Investigation of Anammox Process Using Existing Models and Experimental Approaches

Logos Verlag Berlin

Bibliografische Information der Deutschen Nationalbibliothek

Die Deutsche Nationalbibliothek verzeichnet diese Publikation in der Deutschen Nationalbibliografie; detaillierte bibliografische Daten sind im Internet über http://dnb.d-nb.de abrufbar.

ISBN 978-3-8325-4837-7

Logos Verlag Berlin GmbH
Comeniushof, Gubener Str. 47,
D-10243 Berlin
Tel.:   +49 (0)30 / 42 85 10 90
Fax:   +49 (0)30 / 42 85 10 92
http://www.logos-verlag.de

UNIVERSITÄT
D U I S B U R G
E S S E N

# Optimization and investigation of anammox process using existing models and experimental approaches

A dissertation submitted in partial fulfillment of the requirements for the degree of

Doctor of Engineering (Dr.-Ing.)

University of Duisburg-Essen, Germany

Faculty of Engineering

Department of Urban Water- and Waste Management

**Defended and graduated on 29<sup>th</sup> Nov 2018**

Author:

Mohammad Ebrahim Azari Najaf Abad

Born 23.06.1988

in Mashhad, Iran

Accepted on the recommendation of

1<sup>St</sup> Assessor: Prof. Dr. Martin Denecke (University of Duisburg-Essen)

2<sup>nd</sup> Assessor: Prof. Dr. Susanne Lackner (Technische Universität Darmstadt)

## Abstract

The anaerobic ammonium oxidation (anammox) process has become popular as energy saving, and cost-effective biological nitrogen removal because it shortens the ammonia removal cycle, and directly converts ammonium ($NH_4^+$) to nitrogen gas using nitrite ($NO_2^-$) as an electron acceptor. During recent years, mathematical models explaining interactions between autotrophic and heterotrophic microorganisms including ammonia-oxidizing bacteria (AOB), nitrite-oxidizing bacteria (NOB), anammox bacteria and denitrifiers proved to be substantial for cost reduction, optimization of the system performance and operating conditions. In this work, experimental and mathematical modeling approaches were combined to investigate mechanisms influencing system performance and microbial consortia dynamics in granular and biofilm reactors.

In the first chapter, the 10 years performance of biological treatment for high ammonium removal from a full-scale landfill leachate has been showed. The plant was upgraded combining the activated sludge process followed by granular activated carbon (GAC) biofilm reactor. Based on a long-term analysis, the average total nitrogen removal efficiency of 94 % was achieved for wastewaters with a C: N ratio varying from 1 to 5 kgCOD kgN$^{-1}$. But without the presence of activated carbon reactor, the average of biological removal efficiency for total nitrogen was only 82 % ± 6 % for the activated sludge stage. It means that up to 20 % of the nitrogen in the influent can only be eliminated by microorganisms attached to GAC particles in the form of biofilm. After upgrades of the plant, the energy efficiency showed a reduction in the specific energy demand from 1.6 to less than 0.2 kWh m$^{-3}$. Methanol consumption and sludge production was reduced by 91 % and 96 %, respectively. Fluorescent in situ Hybridization (FISH) was used for microbial diversity analysis on floccular and granular sludge samples. Anammox bacteria and nitrifiers were detected and *Candidatus* Scalindua was found in two forms of flocs and biofilms. Due to stochastic risk assessment based, the treatment criteria were achieved and the combination of GAC biofilm process and activated sludge can be a sought approach to better enrich anammox biomass for full-scale treatment applications to reduce operating costs and promote nutrient removal stability and efficiency.

Second, a mathematical model is proposed and validated for biological nitrogen removal in a granular system to describe independent short-term batch processes under anoxic conditions. The activated sludge model consists of anammox and heterotrophic bacteria using a novel stoichiometric matrix. Identifiability of sensitive biokinetic parameters of the model was assessed with regards to observed concentrations of ammonium-, nitrite-, and nitrate-nitrogen. The Chi-squared function was used for the error estimation and the R-squared index ($R^2$) was used for the regression analysis. The results imply that the model can elucidate the interactions of nitrogen converting bacteria for various feeding characteristics. The calibration results showed satisfactory $R^2$ equal to 0.95 and 0.97 for $NH_4$-N and $NO_2$-N respectively. For validation, model simulations were performed under three varying scenarios and $R^2$ was more than 0.9 so that all forecasted values lied within the 95 % confidence interval. In addition, the estimated physiologic characterization of two dominant anammox species was discussed upon calibration and validation of the model. The maximum specific bacterial growth rates ($\mu_{max}$) for *Candidatus* Brocadia anammoxidans and *Ca.* Scalindua sp. were estimated at 0.0025 $h^{-1}$ and 0.0048 $h^{-1}$ respectively. Decay rate of *Ca.* Brocadia anammoxidans was estimated at 0.0003 $h^{-1}$ which is 15 % higher than decay rate of species belonging to *Ca.* Scalindua.

In the next step, a comprehensive model explaining the roles and functions of soluble microbial products (SMP) and extracellular polymeric substance (EPS) was developed. Therefore, the model for a granular reactor for the leachate treatment discussed in the first chapter was validated by long-term measured data to investigate the mechanisms and drivers influencing biological nitrogen removal and microbial consortia dynamics. The proposed model, based on Activated Sludge Model (ASM1), included anammox, nitrifying and heterotrophic denitrifying bacteria which can attach and grow on GAC particles. Two kinetic descriptions for the model were proposed: with and without SMP and EPS. The model accuracy was checked using recorded total inorganic nitrogen concentrations in the effluent and estimated relative abundance of active bacteria using quantitative FISH (qFISH). The models with and without EPS successfully simulated the relative abundance of the biomass in the biofilm solid matrix, with a high agreement with results obtained from qFISH. Nevertheless, the model was improved after the addition of EPS and SMP kinetics and the real-time production of EPS and SMP can be predicted as well. The models could predict the dynamics of nitrogen transformation well, and a linear relationship between predicted and measured total nitrogen was achieved ($R^2 = 0.66$ for the model with EPS and $R^2 = 0.61$ for the model without EPS). Averagely, the models predicted the relative abundance of 52 % for anammox, 5.2 % for AOB, 2.6 % for NOB and 22 % for heterotrophs. For the model with EPS, the EPS content contributed between 4 to 10 % of total biomass volume. Results suggested that the model with EPS fits better for active relative abundance (the standard error was less than 10 % of observation). The model with EPS and SMP also confirms that the growth and existence of heterotrophs in anammox biofilm systems slightly increases due to including the kinetics of SMP production in the model. During the one-year simulation period, the fractions of autotrophs and EPS in the biomass were almost stable but the fraction of heterotrophs decreased which is correlated with the reduction in nitrogen surface loading (NSL) on the biofilm.

In the last chapter, feasibility of simultaneous nitrification, anammox and denitrification (SNAD) was tested using an integrated fixed-film activated sludge-sequencing batch reactor IFAS-SBR and another SBR mixing activated sludge with granules. During operation of both plants, a regular reduction of temperature was applied, and various aeration regimes were evaluated. IFAS-SBR reached to a stable performance after a short startup period of 25 days and it showed a higher nitrogen removal rate (NRR) compared to another reactor. Highest NRE was able to reach up to 99.9 % based on $NH_4^+$-N concentration and 91.2 % based on total inorganic nitrogen (TIN) concentration. Whereas for NRR, the highest values are 0.14 kgN $m^{-3} d^{-1}$ based on $NH_4^+$-N concentration and 0.15 kgN $m^{-3} d^{-1}$ based on TIN concentration. The optimized range for daily average dissolved oxygen (DO) concentration was determined to be between 0.2 – 0.7 mg $L^{-1}$ and maximum real-time DO at the end of each aeration cycle was set between 1.0 to 1.2 mg $L^{-1}$. Reduction of temperature caused an expected decline of NRE, although a good adaptation was achieved for temperature above 15 °C. For example, after 75 days of the operation, the NRE of 71.1 % was achieved despite the temperature was reduced to 20°C. For microbial identification, 16S ribosomal RNA (16S rRNA) gene sequence analysis and qFISH were applied and AOB was major bacteria group in flocs and anammox bacteria and AOB were found in biofilms attached to the carriers.

**Abstrakt**

Die anaerobe Ammonium-Oxidation (Anammox) Verfahren zur Stickstoffentfernung ist ein kosteneffektiver biologischer Vorgang. Dynamische mathematische Modelle für die Interaktion zwischen heterotrophen und autotrophen Organismen in Biofilmsystemen inkl. Ammoniakoxidierer (AOB), Nitritoxidierer (NOB), Anammox- und denitrifizierende Bakterien sind praktisch und nützlich. In dieser Arbeit wurden experimentelle und mathematische Modellierungsansätze kombiniert, um Mechanismen zu untersuchen, die auf die mikrobielle Populationsdynamik bei Einsatz von Biofilmsystem oder Granularen einwirken. Zuerst wurde die Durchführung einer biologischen Stickstoffbehandlung für eine hohe Ammoniumentfernung aus Deponiesickerwasser mit Kombination des Belebtschlammverfahrens und granuliert Aktivkohle Reaktor gezeigt. Aufgrund der Analyse der von 2006 bis 2015, wurde der mittlere Wirkungsgrad der Stickstoff-Elimination von 94% für Abwässer mit einem C: N-Verhältnis zwischen 1 und 5 kgCOD kg $N^{-1}$ erreicht. Ohne Verwendung von Biofilm auf Aktivkohle-Reaktor betrug der Mittelwert der Wirkungsgrad der biologischen Stickstoff-Elimination ist nur 82% ± 6%. Die Fluoreszenz-in-situ-Hybridisierung (FISH) wurde für die mikrobielle Diversitätsanalyse an Belebtschlamm und Granulat Schlamm verwendet. Anammox Bakterien und AOB wurden nachgewiesen und *Candidatus* Scalindua wurde in beiden Formen von Flocken und Biofilmen gefunden. Nächster Schritt wird ein mathematisches Biofilm Modell für die biologische Stickstoffentfernung in einem Biofilm-Granulat-Reaktor validiert. Es simuliert kurzzeitige Batch-Experimente durch anoxischen Bedingungen. Ein Belebtschlamm Modell Nr. 1 der IWA (ASM 1) besteht aus Anammox- und heterotrophe Mikroorganismen für die Anamox und Denitrifikation mit Entwicklung einer neuen stöchiometrischen Matrix. Die Identifizierbarkeitsanalyse und Sensibilitätsanalyse für biokinetischer Parameter des Modells wurde für Messungen von anorganischen Stickstoffkomponenten (Ammonium-, Nitrit- und Nitrat-Stickstoff) durchgeführt. Für die Modellkalibrierung und -validierung wurden verschiedene Batch-Szenarien entworfen und Das Bestimmtheimmaß ($R^2$) wurde mehr als 0,9 berechnet. Die maximale spezifische Wachstumsrate ($\mu_{max}$) wurden geschätzt: 0,0025 h-1 für *Ca.* Brocadia anammoxidans bzw. 0,0048 h-1 für *Ca.* Scalindua sp. Die Reduktionszeit oder Abklingrate für *Ca.* Brocadia anammoxidans wurde auf 0,0003 h $^{-1}$ geschätzt. Dies ist 15% höher als Arten der Gattung *Ca.* Scalindua. Im nächsten Schritt wurde das Modell erweitert und die Rollen und Funktionen von löslichen mikrobiellen Produkten (SMP) und extrazellulären polymeren Substanzen (EPS) untersucht. Zwei kinetische Beschreibungen für das Modell wurden vorgeschlagen: mit und ohne SMP und EPS. Der Grad der Genauigkeit des Modells wurde anhand von gemessenen anorganischen Gesamt-Stickstoffkonzentrationen im Abwasser und geschätzten relativen Anteils von aktiver mikrobieller Populationen mit der Methode quantitative FISH (qFISH) überprüft. Das Modell wurde nach der Addition von EPS- und SMP verbessert und das Modell kann die dynamische Produktion von EPS und SMP simulieren. Im Durchschnitt geben die Modelle den relativen Anteil von 52% für Anammox, 5,2% für AOB, 2,6% für NOB und 22% für heterotrophe Denitrifikanten. Bei dem Modell mit EPS liegt der EPS-Anteil im festen Biofilm zwischen 4 und 10%. Die Ergebnisse bestätigen, dass das Modell mit EPS mit einem Stichprobenfehler von weniger als 10% besser zur Simulation der aktive relative Anteil passt. Das Modell mit EPS und SMP ergibt höhere Anteil von Heterotrophen in Anammox-Biofilmsystemen aufgrund der Kinetik der SMP-Produktion im Modell (Weil SMP Substrat für die Heterotrophen sein kann). Am Ende wurde der simulatanen Nitrifikation, Anammox und Denitrifikation (SNAD) mit Integration von konventionelle Aktivschlamm-Technologie und Biofilm-Systeme in einem einzigen Sequenzierung-Batch-Reaktor (IFAS-SBR) durchgeführt. Während des Betriebs der Anlagen wurde eine regelmäßige Reduktion der Temperatur durchgeführt, und verschiedene Belüftungsregime wurden bewertet. IFAS-SBR erreichte nach kurzer Anlaufzeit eine stabile Leistung und zeigte im Vergleich zu einem anderen Reaktor eine höhere Stickstoffentfernungsrate (NRR). Der optimierten Konzentration von gelöstem Sauerstoff (DO) wurde zu 0,2 bis 0,7 mg $L^{-1}$ bestimmt und die maximale DO wurde am Ende jedes Belüftungszyklus auf 1,0 bis 1,2 mg L-1 eingestellt. Die Anpassung an die Temperaturreduktion wurde nur für Temperaturen über 15° C erreicht.

**List of publications used in this thesis**

Chapter 2: Azari, M., Walter, U., Rekers, V., Gu, J.D. and Denecke, M., 2017. More than a decade of experience of landfill leachate treatment with a full-scale anammox plant combining activated sludge and activated carbon biofilm. Chemosphere, 174, pp.117-126.

Chapter 3: Azari, M., Lübken, M. and Denecke, M., 2017. Simulation of simultaneous anammox and denitrification for kinetic and physiological characterization of microbial community in a granular biofilm system. Biochemical Engineering Journal, 127, pp.206-216.

Chapter 4: Azari, M., Le, A.V., Lübken, M. and Denecke, M., 2018. Model-based analysis of microbial consortia and microbial products in an anammox biofilm reactor. Water Science and Technology, 77 (7), pp.1951-1959.

Chapter5: Azari, L., Jurnalis, A. and Denecke, M. 2019. The effect of regulation of aeration and temperature on nitrogen removal and microbial community structure in a hybrid sequencing batch reactor, (*in review* at Scientific Reports)

**Other publications as main author or co-author which were cited in this thesis**

Ke, Y., Azari, M., Han, P., Görtz, I., Gu, J.D. and Denecke, M., 2015. Microbial community of nitrogen-converting bacteria in anammox granular sludge. International Biodeterioration & Biodegradation, 103, pp.105-115.

Azari, M.; Le, A.V.; Denecke, M. 2017. Population Dynamic of Microbial Consortia in a Granular Activated Carbon-Assisted Biofilm Reactor: Lessons from Modeling. In Frontiers International Conference on Wastewater Treatment and Modeling (pp. 588-595), Ed. Mannina, G., part of Lecture Notes in Civil Engineering, Vol 4. Springer, Cham, Switzerland

## Acknowledgement

It has been a grateful and fruitful experience and collaboration during the doctoral thesis, which brought me new skills, challenges and opened a broader window towards my scientific career. First, I would like to thank my main supervisor Prof. Martin Denecke who provided me precious advices during this work. I am so honored to have him as my mentor and advisor and I sincerely gratitude him for understanding and thoughtfulness over four years. I really appreciate him for guiding me towards the right path of life and research!

I would like to thank all the researcher and technical staffs in the laboratories coordinated by Department of Urban Water- and Waste Management at University of Duisburg-Essen. Also, would like to thank Dr. Manfred Lübken at Ruhr University Bochum to be my co-advisor through my PhD project by providing me great ideas and mentoring me during my research activities. Besides I acknowledge Prof. Dr. Susanne Lackner at TU Darmstadt to accept to be the reviewer and examiner of my dissertation.

In addition, I would like to acknowledge the support of the German Academic Exchange Service (DAAD) for providing me the NaWaM scholarship to accomplish this doctoral program at University of Duisburg-Essen. I also appreciate the collaboration with the AGR Group and LAMBDA Gesellschaft für Gastechnik mbH, and especially Mr. Volker Rekers for providing me the samples and data. I am also thankful of colleagues at Biofilm Centre, Faculty of Chemistry, University of Duisburg-Essen for providing microscopy facilities and the lab spot to do DNA extraction experiments. I also would like to thank all my colleagues Erika, Asma, Leon, Lukas, Thomas, Danica, and Sonja for their help and supports.

Finally, I want to express from bottom of my heart, my appreciation for my love and my everything „Van" for all her supports, care, and tolerance during my PhD work: She is the person who is always encouraging me and cheering me and standing beside me throughout my career. Thank you!

I would like to also gratitude my family members and I want to dedicate this dissertation to my mother and my father for all their attempts to teach me to persevere and prepare to face the challenges!

## Table of contents

# List of abbreviations

| | |
|---|---|
| AMO | ammonia monooxygenase |
| AMX | anammox bacteria |
| Anammox | anaerobic ammonium oxidation |
| AOB | ammonia oxidizing bacteria |
| ASM | activated sludge model |
| BAP | biomass-associated products |
| BOD | biochemical oxygenic demand |
| BP3 | Benchmark problem 3 |
| CLSM | confocal laser scanning microscopy |
| COD | chemical oxygenic demand |
| DNA | deoxyribonucleic acid |
| DO | dissolved oxygen |
| EDTA | ethylenediaminetetraacetic acid |
| EPS | extracellular polymeric substance |
| FISH | fluorescence in situ hybridization |
| FOV | fields of views |
| GAC | granular activated carbon |
| HAO | hydroxylamine dehydrogenase |
| HRT | hydraulic retention time |
| IFAS | integrated fixed-film activated sludge |
| IKS | internes Kontrollsystem (in English: internal system control) |
| MBBR | moving bed biofilm reactor |
| MBR | membrane bioreactor |
| MLSS | mixed liquor suspended solid |
| MLVSS | mixed liquor volatile suspended solid |
| N/D | nitrification-denitrification |
| NLR | nitrogen loading rate |
| NOB | nitrite oxidizing bacteria |
| NRE | nitrogen removal efficiency |
| NRR | nitrogen removal rate |

| | |
|---|---|
| PN/D | partial nitritation denitrification |
| PN-A | partial nitritation-anammox |
| PVC | polyvinyl chloride |
| qFISH | Quantitative fluorescence in situ hybridization |
| RNA | ribonucleic acid |
| rRNA | ribosomal ribonuleic acid |
| SAA | specific anammox activity |
| SBR | sequencing batch reactor |
| SDS | sodium dodecyl sulphate |
| SAD | simultaneous anaerobic ammonium oxidation and denitrification |
| SBR | sequencing batch reactor |
| SMP | soluble microbial products |
| SNAD | simultaneous partial nitrification, anaerobic ammonium oxidation and denitrification |
| SRBC | submerged rotating biofilm contactors |
| SRT | solid retention time |
| SS | suspended solid |
| TIN | total inorganic nitrogen |
| TN | total nitrogen |
| UAP | utilization-associated products |
| VSS | volatile suspended solid |
| WWTP | wastewater treatment plant |
| ZDA | Zentraldeponie Aachen (In English: central solid waste disposal site and leachate treatment plant in Alsdorf-Warden, Aachen in Germany) |
| ZDE | Zentraldeponie Emscherbruch (In English: central disposal site and leachate treatment plant in Gelsenkirchen along the river Emscher in Germany) |

# List of figures

nitrite ($NO_2^-$), nitric oxide (NO) and nitrous oxide ($N_2O$) reductases to dinitrogen gas ($N_2$) encoded by nar, nir, nor and nos gene clusters, respectively. Enzymes for denitrification of nitrite (B) are: nitrite, nitric oxide and nitrous oxide reductases encoded by nir, nor and nos gene clusters, respectively. Other pathways can be nitrite respiration (nir) and nitric oxide deoxygenation (nod). Consequently, monooxygenases (MO) are enzymes that catalyse the released oxygen atom from $O_2$ into an organic substrate such as alcohols. In anammox process (C), hydrazine ($N_2H_4$) is oxidized to $N_2$.

**List of tables**

# 1 Introduction

## 1.1 Biological nitrogen removal

Nitrogen is an essential component of living organisms and one of the main nutrients limiting life on our planet. As an imbalance within the nitrogen cycle (Fig. 1), the excessive amount of nitrogen which is usually present in the form of ammonium or organic nitrogen have been accumulated in natural ecosystem and have become a big concern toward the end of the 20th century (van Loosdrecht et al., 2015). The environmental impacts associated with ammonia nitrogen include promotion of eutrophication, toxicity to aquatic organisms, and depletion of dissolved oxygen in receiving water bodies because of bacterial oxidation of ammonia to nitrate (Rahimi et al., 2011). Consequently, nitrogen-removing systems were used in many wastewater treatment plants, which were primarily used to remove only organic carbon (Kuypers et al., 2018). The nitrogen removal processes in the wastewater was mostly implemented by physicochemical and biological methods. However, the biological treatment using activated sludge is more effective and economic than physicochemical treatment, thus it has been widely applied (Khin and Annachhatre, 2004).

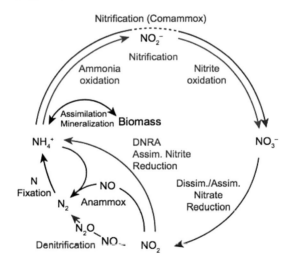

Figure 1: Key processes of biological nitrogen removal (Daims et al., 2016)

### 1.1.1 Conventional nitrification-denitrification

Conventionally, biological nitrogen removal process includes two steps nitrification, and denitrification. The aerobic nitrification is the biological conversion of ammonium to nitrite and nitrate, using oxygen as the electron acceptor. In the first step, the ammonia-oxidizing bacteria (AOB) are responsible for the Eq. 1, in which 1.5 mole of oxygen is used to convert one mole ammonium to nitrite. The bacteria involved are from genera of *Nitrosomonas*, *Nitrosococcus*, *Nitrosospira*, *Nitrosovibrio*, and *Nitrosolobus*. In the second step, nitrite-oxidizing bacteria (NOB) convert each mole of nitrite and a half mole of oxygen to one mole of nitrate (Eq. 2). NOB include genera: *Nitrobacter*, *Nitrospira*, *Nitrospina*, *Nitrococcus*, and *Nitrocystis* (Grady et al., 1999). Later, Daims et al (2015) and van Kessel et al (2015) have also discovered single bacteria, belonging to the genus *Nitrospira*, which are able to catalyze complete nitrification called commamox microorganisms (Daims et al., 2015; van Kessel et al., 2015).

$$NH_4^+ + 1.5O_2 \rightarrow NO_2^- + 2H^+ + 2H_2O \ (\Delta G^{0\prime} = -275kJ \ mol^{-1}) \hspace{2cm} \text{(Eq. 1)}$$

$$NO_2^- + 0.5O_2 \rightarrow NO_3^- \hspace{5cm} \text{(Eq. 2)}$$

The anoxic denitrification is accomplished mostly by heterotrophic denitrifying bacteria, which is commonly found among phylum of *Proteobacteria*, e.g. some of these species belong to genera *Pseudomonas*, *Alcaligenes* and *Paracoccus* (Madigan et al., 2012). This species converts the nitrite and nitrate to the nitrogen gas (Eq. 3 and 4). Denitrification process requires the availability of organic carbon source, such as methanol, ethanol, glucose and acetate (Akunna et al., 1993; Tam et al., 1992).

$$2NO_3^- + 10H^+ + 10e^- \rightarrow N_2 + 2OH^- + 4H_2O \hspace{3cm} \text{(Eq.3)}$$

$$2NO_2^- + 6H^+ + 6e^- \rightarrow N_2 + 2OH^- + 2H_2O \hspace{3cm} \text{(Eq.4)}$$

The conventional nitrification-denitrification is quite costly due to the large amount of oxygen and organic carbon demands for nitrification and denitrification respectively. For example, a study of Gujer and Jenkins in 1975 showed that each gram of ammonium nitrogen needed 2.5 g of dissolved oxygen for nitrification, and another study of McCarty et al 1969 mentioned that one gram of nitrate nitrogen requires 2.47 gram of ethanol for the denitrification process (Gujer and Jenkins, 1975; McCarty et al., 1969). Therefore, several alternative low-cost treatment strategies for nitrogen removal in wastewater have been introduced.

### 1.1.2 Partial nitrification – denitrification (PN/D) process

The PN/D is a biological nitrogen removal (BNR) process via nitrite pathway. Since in the wastewater treatment plant (WWTP), the aeration process has been one of the energy consuming sections (Jonasson and Ulf Jeppsson, 2007), to save energy, the partial nitrification of ammonia to nitrite (Eq. 1.) by repression of NOB combined with the second step of denitrification (Eq. 4.) has been proposed. The PN/D process is highly applicable for wastewater with high ammonium concentration ($\geq 500$ mgN L$^{-1}$). The most challenging issue of this method is to achieve the stable partial nitrification, where parameters such as pH, temperature, dissolved oxygen, and hydraulic retention time, are controlled in a system. The required dissolved oxygen (DO) concentration by nitritation by AOB is favorably low. However, the nitritation rate reduces in lower temperatures (Fig. 2), then it lowers chemical oxygen demand (COD) efficiencies and causes sludge bulking. Nevertheless, nitritation could convert to a complete nitrification when DO is high ($>2$ mg L$^{-1}$). Therefore, a suitable DO concentration is in the range $0.3 - 2.5$ mg L$^{-1}$ for nitritation (Stijn et al., 2004). Compared to conventional nitrification and denitrification via nitrate, PN/D process requires less oxygen to oxidize ammonium to nitrite, less organic carbon for the denitrification of nitrite and achieves a lower sludge production (She et al., 2016).

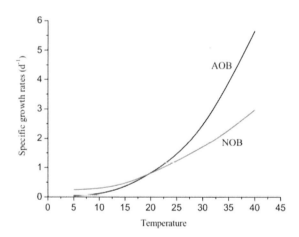

Figure 2: Effect of temperature on growth rate of AOB and NOB (Zhu et al., 2008).

Temperature influences the growth of NOB and ammonia oxidizing bacteria (AOB). If the temperature remains low (<15° C), NOB will be dominating over AOB (e.g. *Nitrosomonas sp.*). Whereas at higher temperature (>20° C), AOB will outcompete NOB (Brouwer et al., 1996). NOB also needs a longer solid retention time (SRT) at temperatures above 20 (see Fig. 2), while the trend reverses for AOB.

### 1.1.3 Discovery of anaerobic ammonium oxidation (anammox) process

From 1940s to 1970s, the ammonium accumulated in anoxic water body was found less than expectation, and thermodynamic calculation showed the free energy for the reaction of anaerobically oxidizing ammonium to nitrogen gas is favorable. The hypothesis of a missing microbe which is able to convert ammonium to nitrogen gas in one step was considered (Van Niftrik and Jetten, 2012). However, until 1990s this process was postulated in a WWTP as ammonium could be converted to nitrogen gas by using expense nitrate from an anoxic fluidize-bed reactor (Eq. 5) (Mulder et al., 1995). But it was later proved that anammox is a biologically mediated process and dinitrogen gas is the end product and nitrite as the preferred electron acceptor instead of nitrate, hence the anammox reaction was presented as Eq. 6 (Van de Graaf et al., 1995; Van de Graaf et al., 1997). *Planctomycetes* is the most dominant phylum in anammox process. The bacteria are majorly autotrophs consuming ammonium as their energy source (electron donor) and dissolved $CO_2$ and $HCO_3$ is used for biosynthesis (Suneethi et al., 2014).

$$5NH_4^+ + 3NO_3^- \rightarrow 4N_2 + 2H^+ + 9H_2O \qquad \text{(Eq. 5)}$$

$$NH_4^+ + 1.32NO_2^- + 0.066HCO_3^- + 0.13H^+ \rightarrow 1.02N_2 + 0.26NO_3^- + 0.066CH_2O_{0.5}N_{0.15} + 2.03$$
$$H_2O \ (\Delta G^{0'} = -357kJ \ mol^{-1}) \qquad \text{(Eq. 6)}$$

Later in 2014 the revised stoichiometry of the anammox process was given by performing kinetic batch experiments and analyzing the constitute elements (Eq. 7). This revision (Eq. 7) claimed that the release of nitrate during the anammox process is 38 % less than what it was assumed before in Eq. 6 (Lotti et al., 2014).

$$1 \ NH_4^+ + 1.146 \ NO_2^- + 0.071 \ HCO_3^- + 0.057 \ H^+ \rightarrow 0.986 \ N_2 + 0.161 \ NO_3^- + 2.02 \ H_2O +$$
$$0.071 \ CH_{1.74}O_{0.31}N_{0.2} \qquad \text{(Eq. 7)}$$

Since its discovery, it is known as a promising technology to help WWTPs to achieve energy recovery (Van Loosdrecht et al., 2014). This requires no addition of an external carbon source, less excess sludge production (~90 %) and no aeration and no direct contribution to greenhouse gases, i.e. $N_2O$ emission (Fux et al., 2004).

Besides, due to the anoxic condition, DO concentrations need to be preferably less than ~0,2 mg $L^{-1}$. Otherwise, the anammox activity will be severely reduced (Strous & Jetten, 1997). Also, despite $NO_2^-$ is required for anammox process but high amount of $NO_2^-$ can be another inhibitor in anammox activity due to its nitritation capacity, which requires oxygen for oxidation. The highest anaerobic ammonia-oxidizing activity was achieved at 25 ppm $NO_2$ in the headspace while $NO_2$ more than 50 ppm deteriorated anammox activity (Schmidt & Bock, 1997). Whereas, $NH_4^+$ and $NO_3^-$ concentration cause only few inhibition (Marc Strous & Jetten, 1997). Percentage of $NH_4^+$ and $NO_3^-$ and possible free ammonia in wastewater is also influenced by pH, which can affect anammox performance. By far, the optimal range of pH in which anammox process is reliably working, is approximately considered between 7.7 – 8.3 while the most preferable range can be as ~ 7.5 to 7.7 (Strous and Jetten, 1997; Tomaszewski et al., 2017). Additionally, anammox activity can be expected at 5° – 45° C. However, the reaction rate will decline rapidly when temperature goes below 15° C or over 40° C. The optimal temperature is around 30° C as shown in Fig. 3 (Lotti et al., 2015).

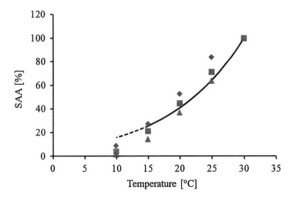

Figure 3: specific anammox activity (SAA) measured at different temperatures after normalization at 30 °C. The solid line (continuous for the temperature range 15° - 30°C and dashed for 10° - 15 °C) The graphs is based on the Arrhenius equation using 30°C as reference temperature and an activation energy (Ea) equal to 65.7 kJ.mol $L^{-1}$ (Lotti et al., 2015).

### 1.1.4 Partial nitritation-anammox (PN/A)

Partial nitritation by autotrophic AOB (such as *Nitrosomonas*) in oxygen limited conditions allow to remove ammonium over nitrite (Eq. 8), and in the next stage anammox bacteria convert ammonium with the produced nitrite in step one to nitrogen gas and trace amount of nitrate (Eq. 9). At the beginning, this process was implemented separately in two stages, and in 2011 the PN/A was introduced into one single reactor (Third et al., 2001).

$$NH_4^+ + 1.5O_2 \rightarrow NO_2^- + 2H^+ + 2H_2O \qquad \text{(Eq. 8)}$$

$$NH4^+ + 1.3NO_2 \rightarrow 1.02N_2 + 0.26NO_3^- + 2H_2O \qquad \text{(Eq. 9)}$$

The PN/A process, also called deammonification, is one of the most effective and economic for nitrogen removal, specially, with the influent containing high ammonium concentration but less organic carbon. Compared to the conventional technique, PN/A requires no addition of an external carbon source, less excess sludge production ($\sim$90 %) and no aeration and less energy due to aeration ($\sim$63 %) (Van der Star et al., 2007; Tang et al., 2011; Lackner et al., 2015).

Over more than a century of the development of nitrogen removal technology, the recent nitrogen removal processes are compared with conventional method in Table 1 and Fig. 4. To reduce the total expenses and maximize the removal efficiency, anammox and partial nitrification-anammox processes are prominent solutions for enhance nitrogen removal. Anammox process has been successfully applied to treat high-strength ammonium-containing wastewaters such as sludge digester effluent, leachate, or industrial wastewater.

Table 1: A comparison of alternative nitrogen removal processes and conventional nitrification-denitrification (adapted from Jetten et al., 2002)

| System | Conventional nitrification-denitrification | Partial nitrification-denitrification | Partial nitritation-anammox or deammonification |
|---|---|---|---|
| Number of reactors | 1 or 2 | 1 | 1 |
| Feed | Wastewater | Wastewater | Wastewater |
| Discharge | $NO_2^-$, $NO_3^-$, $N_2$ | $NH_4^+$, $NO_2^-$ | $NO_3^-$, $N_2$ |
| Conditions | Oxic, anoxic | Oxic | Oxygen limited |
| Oxygen requirement | High | Low | None |
| pH control | Yes | None | None |
| Biomass retention | None | None | Yes |
| COD requirement | Yes | None | None |
| Sludge production | High | Low | Low |
| Major involving bacteria | AOB, NOB and various heterotrophs | AOB and various heterotrophs | AOB and anammox bacteria |

Deammonification can occur in two separated tanks, i.e. the combined SHARON/Anammox process reported by Van Dongen et al. (2001) or in one system in relatively limited oxygen condition. Such one reactor systems are reported as various acronyms, e.g. oxygen-limited autotrophic nitrification–denitrification (OLAND), completely autotrophic nitrogen-removal over nitrite (CANON), partial nitritation anaerobic ammonium oxidation process (PNAA), single-stage nitrogen removal using anammox and partial nitritation (SNAP) and DEMON deammonification technology (Van Hulle et al., 2010; Takekawa et al., 2014; Ali and Okabe, 2015).

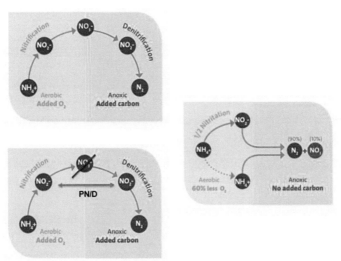

Figure 4: A schematic comparison of alternative low-cost nitrogen removal processes and conventional nitrification-denitrification

### 1.1.5 Simultaneous partial nitrification anammox and denitrification (SNAD)

Due to the possibility of nitrite and nitrate accumulation during deammonification, applying a successful combination of conventional denitrification and anammox process can be useful. SNAD process is a complex anammox-based biological process with different populations in oxygen-limited conditions. It combines partial nitrification, anammox process and denitrification as major processes occurring simultaneously in one single reactor under oxygen limitation (Lan et al., 2011). In addition, recent studies have shown that low amount of organic compounds (limitation of organic carbon) in deammonification systems can improve the overall nitrogen removal through the partial heterotrophic denitrification, which reduces nitrate to nitrite and more nitrite can be taken up by anammox organisms which can lead to increase in nitrogen removal efficiency (Mozumder et al., 2014). Thus, the conditions are micro-aerobic for partial nitritation and anoxic for anammox and denitrification. SNAD process has various advantages compared to two-stage systems and other conventional biological treatments since it has lower capital costs, no external carbon sources, lower sludge productions and lower energy and oxygen requirements. A key factor for the development of the SNAD process is a better understanding of both the biological processes involved in the nitrogen removal and the numerous microbial interactions using microbiological and modeling tools (Langone, 2013). Another variant in anoxic conditions (no molecular oxygen) is called simultaneous anammox and denitrification (SAD) (Li et al, 2016).

### 1.1.6 Anammox and dissimilatory nitrate reduction to ammonium (DNRA)

It is recently found that anammox bacteria are capable of dissimilatory nitrate reduction to ammonium (DNRA) with nitrite as intermediate using volatile fatty acids (VFA) as electron donor (organotrophic activity) (Castro-Barros et al., 2017). DNRA by anammox bacteria was reported in two ways:

(i)    converting the nitrate to nitrite only by partial DNRA. Therefore, combining the nitrite formed with external present ammonium forms nitrogen gas from the anammox reaction. The main challenge to establish (partial) DNRA by anammox bacteria is their out-competition by heterotrophic bacteria, which compete for nitrate and acetate (for heterotrophic denitrification) and which have a much faster growth rate.

(ii)   reducing part of the nitrate to the intermediate nitrite and one part to ammonium as products of DNRA which is called full DNRA. This process might be an unfavorable pathway for an effective ammonium removal (Kartal et al., 2007).

### 1.2    Influence of solid retention time for nitrogen removal

Solid retention time (SRT) is the average time that the activated sludge solids are in the system and an important operating design parameter. SRT is determined by dividing the mass of solids in the reactor by the solids removed daily via the effluent and by wasting for process control. SRT is related to the reactor volume, production of solids, oxygen consumption and other operational variables of the process. It is also related with HRT, which expresses in days or hours. HRT is a measure of the average length of time that a soluble compound remains in a constructed bioreactor. The typical SRT values in activated sludge system is 4-10 days, while for HRT 6-8 hours (<0.3 days). For anammox-based nitrogen removal technologies, appropriate biomass retention in reactors is a crucial factor for the accurate operation due to the slow growth rate of anammox planctomycetes (Fernández et al., 2008). At least two different approaches were studied and compared to improve biomass retention by minimizing wash-out events: (1) formation of granular biomass (can be called granular sludge or granular biofilm) (2) In biofilm systems where immobilized cell technologies are used.

### 1.2.1 Biofilm-based systems

In such systems either process with internal fixed media for attached growth of biofilms or dispersed carrier material for biofilm formation can be applied. Some of most common immobilized cell technologies are summarized in Fig. 5. Regarding the packed bed reactors, the advantage is the ability to increase the loading on an existing plant without increasing solids load on existing secondary clarifiers, as most of the biomass is retained in the reactor and biomass concentration in terms of mixed liquor suspended solids (MLSS) which is the concentration of suspended solids in the mixed liquor, can be reached up to 5,000 – 9,000 mg $L^{-1}$ based on the results with full-scale and pilot-scale test (Tchobanoglous et al., 2003). Moving bed biofilm reactors (MBBRs) are a typical biofilm reactor developed in Norway during 1980s and early 1990 where some typical suspended packing materials or carriers, which are from sponge or foam pads such as Captor® and Linpor®, are applied for attached film growth. These sponge carriers will be placed in the bioreactor in a free-floating way and its volume can account for 20-30 % of the reactor volume. Another type of carrier is plastic cylindrical packing e.g. from polyethylene. A Norwegian company created another common cylindrical packing material, Kaldnes Miljøteknologi, during the development of MBBRs (Chrispim and Nolasco, 2017). The process consists of adding small cylindrical-shaped polyethylene carrier elements to support biofilm growth. The small cylinders (K1) are about 10 mm in diameter and 7 mm in height with a cross inside the cylinder and longitudinal fins on the outside. The bigger one is K3 which is about 240 mm in diameter and 100 mm in height. The Kaldnes cylinder could be filling 25-50 % of the tank volume. The specific surface area of this carrier is about 500 $m^2$ $m^{-3}$ of bulk carrier volume. MBBR does not require any return activated-sludge flow or backwashing and also provides an advantage for plant upgrading by reducing the solids loading on existing clarifiers (Rusten et al., 2000). A list of some commercial biofilm carriers for the MBBR/ IFAS application is given in Table 2. The suspended packing also facilitates high oxygen and nutrient transfers in reactors (Ye et al., 2009), which is why this type was used in this experiment. One hybrid form of MBBR technology is IFAS (Sriwiriyarat et al., 2008). IFAS is the integrated fixed film activated sludge process including two fundamental biological treatment processes, fixed-biofilm technology using carriers and suspended growth technology (conventional activated sludge in the form of floccular biomass and/or planktonic cells), together into one hybrid system. Operating the IFAS can be more flexible in a sequencing batch reactor (SBR) since no clarifier is required for a SBR because the sedimentation process will take place in the main reactor (U.S.Environmental Protection Agency, 2015).

Table 2: Examples of the six common commercial biofilm carriers for the MBBR and IFAS

| | Supplier | Name | Specific area | Dimensions (mm) | Material | Image |
|---|---|---|---|---|---|---|
| 1 | Veolia Inc. | AnoxKaldnes™ K4 (MBBR/IFAS) | 800 m² m⁻³ | 4 mm; 25 mm (cylindrical) | PE-Material | |
| 2 | Veolia Inc. | AnoxKaldnes™ K5 (MBBR/IFAS) | 800 m² m⁻³ | 3.5 mm; 25 mm (cylindrical) | PE-Material | |
| 3 | Veolia Inc. | BiofilmChipM | 1200 m² m⁻³⁻ | 2.2 mm; 48 mm (Round) | PE-Material | |
| 4 | Veolia Inc. | ANOXKALSNES™ Z-MBBR | >1200 m² m⁻³ | 3D printing | PE-Material | |
| 5 | Multi Umwelttechnologie AG | Mutag BiofilmChip | 4000 m² m⁻³ | 1.1 mm; 25 mm (Round) | PE Neuware | |
| 6 | Biofilm-Tech GmbH | LEVAPOR | >2500 m² m⁻³ | 7 mm; 20 mm; 20 mm (Cubic) | Polyurethane coated with activated carbon | |

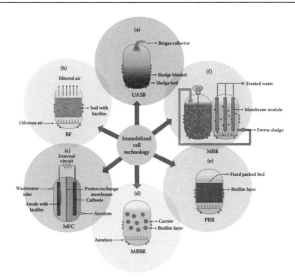

Figure 5: Overview of some common types of bioreactors with biofilm technologies for wastewater treatment: (a) Upflow anaerobic sludge blanket (UASB) bioreactor; (b) biofilter (BF); (c) microbial fuel cell (MFC); (d) moving bed biofilm reactor (MBBR); (e) packed (fixed) bed bioreactor (PBB); (f) membrane biofilm reactors (MMBfR). (ElMekawy et al., 2016)

### 1.2.2 Granular systems

Microbial granular systems are a special application of biofilm-based technologies. Granules (e.g. less than 1 to 10 mm in diameter) are small agglomerations of microorganisms that, because of their weight, resist being washed out (Fig. 6). The granular sludge can be formed in an anoxic or microoxic environment (appropriate for anammox-based processes) or can be aerobic granules (e.g. NEREDA technology). The aerobic granular sludge is known of its excellent settlement's ability; dense and strong microbial structure, high biomass retention; ability to withstand at high organic loading; tolerance to toxicity (Adav et al., 2008). Developing and enrichment of the granulated sludge may take several months. To effectively enrich the slow growing organisms, such as methanogen, anammox and nitrifiers, granular sludge is one of the options (Nicolella et al., 2000). It provides a long SRT and has an internal anoxic zone, which could promote the growth of anammox bacteria (Winkler et al., 2012). A high removal load of pollutants and salinity and high resistance to the influent fluctuation could be achieved by granular sludge system (Carvajal-Arroyo et al., 2013). A special granular sludge type consists of biofilm-covered biomass on granular activated carbon as nucleus (Ke et al., 2015).

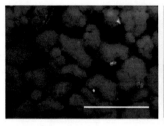 

Figure 6: Anammox-rich granular sludge with different sizes and shapes taken from UASB (bar 1 cm). (Ni and Yang, 2014; Ni and Meng; 2011)

Bulking sludge takes place when readily biodegradable COD is removed under the conditions where strong substrate gradients occur over the sludge floc. Thus, it was realized that granules should form when those conditions are minimized (Beun et al., 1999). It was already hypothesized that biofilm morphology depends on the ratio between substrate transport rate and biomass growth (van Loosdrecht et al., 1995 & 2002). It means not only that minimizing substrate gradients over the sludge floc will improve the SVI, but also selection for slow growing bacteria will improve SVI as well.

## 1.3    Eco-physiology, kinetics, and niche partitioning of anammox species

Anammox bacteria are coccus shaped with a diameter of ~1 μm. Anammox bacteria are phylogenetically from the phylum of *Planctomycetes*. Different from other bacteria, anammox have an intracytoplasmic membrane, which compartmentalizes the cells (Fig. 7). The anammoxosme (ribosome free compartment) is in the innermost of the cell, which occupies the largest volume, and being bounded by an anammoxosome membrane. The riboplasm is the middle compartment which includes ribosome and nucleoid, and the outermost compartment is the paryphoplasm (Van Niftrik and Jetten, 2012).

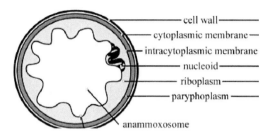

Figure 7: Cellular compartmentalization of *Planctomycetes* (Lindsay et al., 2001)

Unfortunately, anammox organisms have been difficult to cultivate (Egli et al., 2001). Considering on this fact, reactors containing granular sludge were considered to be suitable for the enrichment of anammox bacteria with low growth rates (Tran et al., 2006), as anaerobic *Planctomycetes* have been found in anaerobic granular sludge (Sekiguchi et al., 1998).

The occurrence of anammox has been found in various anoxic environments, such as marine, fresh water and terrestrial ecosystems (Ali and Okabe, 2015). Anammox is difficult to grow and maintain in anoxic pure culture, but their monospecies was enriched in laboratory scale by using different techniques, for example sequencing batch reactor (SBR), membrane bioreactor (MBR) and up-flow column reactor (Ali and Okabe, 2015; Van Niftrik and Jetten, 2012). So far, at least six confirmed *"Candidatus"* anammox bacterial genera have been enriched from wastewater treatment facilities and freshwater environments (*Ca.* Brocadia, Kuenenia, Jettenia, Anammoxoglobus and Anammoxomicrobium), as well as at least one genus from marine environments (*Ca.* Scalindua) (Fig. 8). Recently at least one report proved the existence of *Ca.* Scalindua with other anammox genera in a full scale ammonia-rich leachate treatment plant in UK and Germany respectively (Schmid et al., 2003; Ke et al., 2015; Azari et al., 2017a) as well as one lab-scale UASB for nitrogen removal from waste brine inoculated with marine bacteria (Yokota et al., 2018). Remarkable information of 16S rRNA gene sequences, proteomes and physiological characteristic of *Ca.* Scalindua have highlighted the growth of *Ca.* Scalindua in high salinity conditions, and the activity can be inhibited under the absence of salinity (Sonthiphand et al., 2014).

Since anammox is introduced as an effective alternative nitrogen removal technology, many studies about their physiological and kinetic characteristics have been investigated. Anammox bacteria grow at temperature from 20 to 40 °C, and the optimum pH range is

(13)

between 6.5 to 8.3. The anammox *Ca.* K. stuttgartiensis was the most active at pH 8 and 37 °C, which removal efficiency could reach 26.5 nmol $N_2$. mg protein$^{-1}$ min$^{-1}$. The specific growth rate ($\mu$) of anammox is simulated using the Monod equation (Eq. 10). The maximum growth rate ($\mu_{max}$) and the half saturation constant ($K_S$) on ammonium and nitrite which are also highly depended on other inhibited factors, such as oxygen, 2,4-dinitrophenol, carbonyl cyanide m-chlorophenylhydrazone, acetylene, $HgCl_2$ and etc. (Jetten et al., 1998).

$$\mu = \mu_{max} \frac{S}{K_s + S}$$ (Eq. 10)

Which $\mu$ is the specific growth rate; $\mu_{max}$ is the maximum specific growth rate; S is the concentration of the limiting substrate and $K_S$ is the apparent half-saturation constant. Summary of physiological characteristics of different anammox species are listed in Table 3.

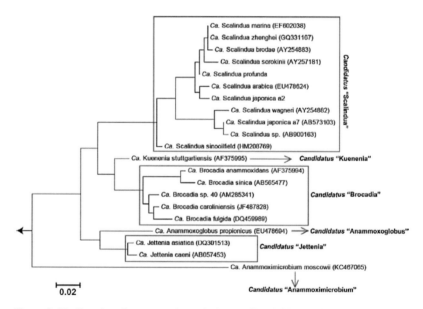

Figure 8: Biodiversity of anammox bacteria from (Ali and Okabe, 2015). Scale bar indicates 2 % sequence divergence

First the doubling time of *Ca.* Brocadia anammoxidans was discovered as being 11 days. However, smaller doubling times were determined for other species of anammox bacteria, such as *Ca.* Kuenenia stuttgartiensis (8.3–11 days), *Ca.* Brocadia sinica (7 days) and *Ca.* Brocadia sp. 40 (2.1 days). Some authors have reported much lower growth rates for *Ca.* Anammoxomicrobium moscowii (doubling time = 32 days) and *Ca.* Jettenia moscovienalis (28 days).

Table 3: Physiological characteristics of *Ca.* Jettenia caeni, *Ca.* Brocadia sinica, *Ca.* Brocadia anammoxidans, *Ca.* Kuenenia stuttgartiensis and *Ca.* Scaclindua sp. ( Zhang et al., 2017)

| Parameter | *Ca.* Jettenia caeni | *Ca.* Brocadia sinica | *Ca.* Brocadia anammoxidans | *Ca.* Kuenenia stuttgartiensis | *Ca.* Scalindua sp. |
|---|---|---|---|---|---|
| Growth temp (°C) | 20-42.5 | 25-45 | 20-43 | 25-37 | 10-30 |
| Growth pH | 6.5-8.5 | 7.0-8.8 | 6.7-8.3 | 6.5-9.0 | 6.0-8.5 |
| Biomass yield (gCOD gN$^{-1}$) | 0.128 | 0.144 | 0.16 | n.d | 0.068 |
| $\mu_{max}$ (d$^{-1}$) | 0.048 | 0.0984 | 0.0648 | 0.0624 -0.084 | 0.048 – 0.17 |
| Affinity ($K_S$) | | | | | |
| NH$_4^+$ (gN m$^{-3}$) | 0.24±0.06 | 0.39±0.056 | 0.07 | n.d | 0.042 |
| NO$_2^-$ (gN m$^{-3}$) | 0.49±0.013 | 0.48±0.3 | 0.07 | 0.003-0.042 | 0.0063 |
| Tolerance | | | | | |
| NO$_2^-$ (gN m$^{-3}$) | 154 | < 224 | 98 | 185.5 | 105 |
| NH$_4^+$ (gN m$^{-3}$) | > 280 | n.d | n.d | n.d | > 224 |
| Sulfide (gS. m$^{-3}$) | 17.28 | n.d | 3.52 | 0.32-9.6 | n.d |

## 1.4 Application of anammox: challenges and solutions

By the end of 2014, more than 114 full-scale anammox-based plants came into being around the world (Ali et al., 2015; Lackner et al., 2014) (Fig. 9). The partial nitrification/anammox (PN/A) processes, which were described have been widely installed in most of the anammox plants. At the beginning PN/A is implemented in two reactors, the partial nitrification takes place first, following anammox reactor commercially known as SHARON/ANAMMOX. However, PN/A has been designed in one single reactor, applying different technologies such as moving bed biofilm reactor (MBBR), granular sludge processes and sequencing batch reactor (SBR). The review from Lackner et al., (2014) has revealed that SBR has been the most applied technology, which has contributed to more than 50 % over total PN/A plants (Lackner et al., 2014).

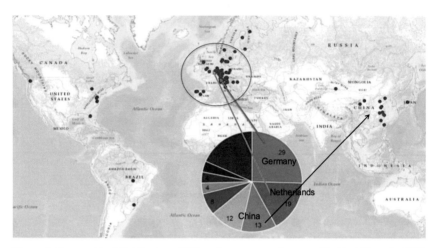

Figure 9: Geographical distribution of full-scale anammox plants worldwide (Ali et al., 2015)

There are reports highlighting the start-up, challenges, and operational troubles of full-scale anammox plants. The major challenges of anammox plants are listed below:

The process requires keeping SRT high enough to reach accumulation of anammox bacteria and selective retention of ammonia-oxidizing bacteria AOB over NOB. This will lead to a longer start-up period (Lackner et al., 2014). But mainstream application of anammox process for influent with lower nitrogen concentration, high COD/N ratio and lower ambient temperatures is highly restricted (Ali and Okabe, 2015).

Accumulation of inorganic nitrogen substances ($NH_4^+$, $NO_2^-$ and $NO_3^-$). For instance, the nitrite is known as inhibitor for the growth of anammox. The accumulation time of nitrite was reported up to one week, and reasons could be a disturbance of anammox community or over aeration, which allowed AOB to grow faster. Or in the case of accumulating nitrite, which is not an inhibited factor but an increase in nitrate signals that the microbial community is unbalanced and too many nitrite oxidizing bacteria have accumulated. (Lackner et al., 2014). Deammonification process produce nitrate in the reaction, which leads to a nitrogen reload of the wastewater and restrict the total nitrogen removal rate to maximal 90 %.

Besides, other issues facing during operation of full-scale anammox were mechanical failures (pump failure, mixing problem and solids concentration); inhibiting conditions, foaming, scaling, settling and solid biomass loss (Lackner et al., 2014).

To address the first issue, DEMON configuration is a successful example for the operation of SBR for PN/A process by DEMON GmbH in Switzerland, which applies a patented pH-based feed control system. Besides in a WWTP in Austria separation of AOB, NOB and anammox bacteria can be handled by a hydrocyclone and washing out NOB can be effectively achieved by controlling of the selected SRT of AOB and anammox bacteria. The NOB in form of smaller flocs are washed out, and anammox with slower growing rate can grow in form of biofilms or larger aggregates. Also, Eawag group (Zürich, Switzerland) also developed another SBR technology, which $NH_4^+$ is under the control by a sensor. The feeding procedure is implemented at the start of each cycle, or during aeration (Lackner et al., 2014). Another significant approach is using a biofilm-based technology or a granular technologies (Vlaeminck et al., 2012) to increase the total SRT. This is because biofilms can sustain microorganisms with very different growth kinetics due to the undefined SRT and their distinct substrate gradients. Examples are the ANITA™Mox processes in a MBBR (Moving Bed Biofilm Reactor) by VeoliaWater Technologies and DeAmmon technology by Purac AB in Sweden. The two steps are taking place in a one-stage biofilm process in different layers of the biofilm grown on plastic media: nitritation (aerobic) in the outer layer of the biofilm, anammox (anoxic) in the inner layer of the biofilm.

## 1.5 Microbial biopolymer secretion and soluble products

In biological wastewater treatment plants, the microbial communities are aggregated in different forms, such as sludge flocs, biofilms, and granules. In term of microbial bioproducts within biofilms systems, there are two different components, which are extracellular polymeric substances (EPS) and soluble microbial products (SMP). A review from Sheng et al (2010) defined EPS as high-molecular-weight polymers, which are produced from cells secretions, lysis and hydrolysis (Sheng et al., 2010). In another approach, Namkung and Rittman defined the soluble microbial products as products of cell lysis through cell's membrane which can be lost during the synthesis or be excreted for other purposes (Namkung and Rittmann, 1986). The soluble microbial products can be produced during the growth and decay of microorganism, hence latter would influence the quality of the effluent. To avoid any misleading in definition of two concepts of EPS and SMP in biofilm, this chapter would like to elucidate in detail the EPS and SMP approaches in wastewater treatment systems, and finally elaborate a utilized hypothesis, which will be adapted for this study. Moreover, an in-depth study of EPS and SMP formations will upgrade the efficiency of WWTs by optimization operational parameters.

### 1.5.1 Extracellular polymeric substances (EPS)

The substances adhering to cells and surfaces are known as extracellular polymeric substances (EPS). The complex interaction among EPS and cells form a net-like structure, which protect cells from desiccation and effect of toxic substances (Wingender et al., 1999). EPS can be classified into soluble EPS and bound EPS including tightly bound (TB) in the inner layer and loosely bound (LB) in the outer layer (Fig. 10). But the soluble EPS bind weakly with cells and can be separated by centrifugation. (Laspidou and Rittmann, 2002a,b; Nielsen and Jahn, 1999). Most of the studies have focused on the bound EPS, and very limited information about the soluble EPS, though they have crucial effect on the microbial activity and surface characteristic of sludge.

In case of anaerobic granular sludge, most of the EPS concentrate in the outer layer, and smaller portion distribute throught the granualar depth. The aerobic granular was reported EPS content mostly found in inner layer, which was four time greather than their outerlayer (Wang et al., 2005). The EPS structure composition are mostly polysaccharides and proteins, but also include other macro-molecules such as nucleic acids, lipids and humic substances. The quantification of each composition and their distribution within biofilm highly depends on microbial aggregate types, structures and origins. The quantification of EPS in wastewater system drives the focus on the production of EPS from living biomass, and assume a part of EPS – soluble EPS is associated with the bulk liquid (Laspidou and Rittmann, 2002a,b). However, this approach seems underestimated the biofilm detachment and hydrolysation of EPS, which a part of it might decay to the inert biomass. The EPS studies have been employed serveral techniques for quantification and in situ charaterization. For instance, to observe and quantify the spatial distribution of bound EPS, the EPS can be stained by flourescent dyes or lectins (Sheng et al., 2010). However, regarding to observe of soluble EPS and its effect to the effluent still require more investigation.

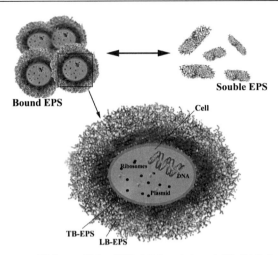

Figure 10: EPS structure (Nielsen and Jahn, 1999): LB: loosely bound, TB: tightly bound

## 1.5.2 Soluble microbial products (SMPs)

Investigation of soluble organics in wastewater treatment system showed that soluble organics were not present in the influent but produced during biological treatment processes. Such compounds were formed during bacterial endogenous decay (cell lysis products) and substrate metabolism, which were found in biological effluents. This leads to overestimation of oxygenic demand (COD) and biological oxygenic demand (BOD) in wastewater treatment output. Moreover, SMP was found to cause internal membrane fouling in membrane bioreactors (Mesquita et al., 2010). Thus, in order to minimize impacts of SMP in biological treatment system, there were several studies Barker and Stuckey, 1999; Namkung and Rittmann, 1986; Rittmann et al., 1987) which focused on characterization, formulation and kinetic modeling of SMP products.

The kinetics formation of SMP in steady-state biofilm reactor was studied primarily in Namkung and Rittmann (1986), which hypothesized two types of SMP. Utilization - associated products (UAP), which is formed during the growth of living organism, and biomass – associated products (BAP) as a part of decay of biomass and formed by EPS hydrolysis process. In biofilm rectors the formation of UAP was more important than BAP for the experimental conditions (Namkung and Rittmann, 1986). Many studies focused on SMP in aerobic systems, but handful of them worked in anaerobic plants. A review of Barker and Stuckey (1999) compared the production of SMP in aerobic and anaerobic systems. The SMP in anaerobic systems contributed approximately 0.2 to 2.5 % of total microbial matters, while in aerobic plants fed with glucose and phenol their values were $3.1 \pm 4.0$ % and $14.7 \pm 3.7$ % respectively. A comparison of one step aerobics treatment and anaerobic treatment showed that COD producing from anaerobic stage was much lower than the UAP in aerobic treatment (Barker and Stuckey, 1999).

It is also essential to assess the characteristics of SMP e.g. molecular weight distribution, biodegradability, and toxicity, to optimize the system performance. The molecular weight range of SMP varied from smaller to 0.5 kDa to greater than 50 kDa. The higher sludge retention time was (>15 days for anaerobic and > 2 days for aerobic systems), the higher molecular weight of SMP was found. The kinetic of biodegradability of SMP was slower than other simple and lightweight substances (Barker and Stuckey, 1999). Regarding SMP removal methods, different advance treatment techniques of SMP were investigated, such as granular activated carbon (GAC), synthetic resin adsorption, ozonation, oxidation, coagulation and break point chlorination (Barker and Stuckey, 1999). But GAC adsorption is the best removal efficient, which in a study of Parkin & McCarty (1975) the GAC could remove up to 85 % of SMP (Parkin and McCarty, 1975). However, some studies of SMP contribution to microbial communities have shown that some of SMP had chelating properties, which was able to uptake metal nutrients and/or protect from toxicity. Another contribution of SMP was related to selection of microbial community in the system and possible contribution into quorum sensing (Mesquita et al., 2010). Since the precise definition of SMP has been open to debate, many works are still required to be done to fully understand their contribution.

### 1.5.3   Utilized theory for EPS and SMP in wastewater treatment

The EPS and SMP are similar as self-products of microorganism for energy generation and biomass synthesis, but a key issue to differentiate two products is solubility. While a small amount of EPS is soluble, and the rest is associated in solid matrix, the SMP is completely soluble. A critical review of (Laspidou and Rittmann, 2002b) proposed the utilized theory, that EPS, SMP, active biomass and inert were involved in one model (Fig. 11).

First, the theory assumes that EPS and SMP are identical. In such complex multispecies wastewater treatment, it is not important from which origin substrate the products are formed, and the intermediate products of incomplete degradation are neglected.

Second, the presence of SMP in liquid phase is degraded and served as electron donor sources for cell synthesis, as the growth of cell, EPS produces.

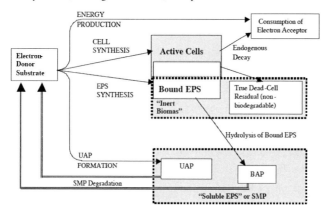

Figure 11: Schematic representation of the unified model for active biomass, EPS, SMP, and inert biomass (Laspidou and Rittmann, 2002b)

Third, the soluble EPS is the same as UAP whose kinetics is captured by the formation kinetic of UAP. The bound EPS is the major proportion of EPS, and be hydrolyzed to BAP. Finally, active biomass (living cells and bound EPS) decay endogenously to true dead-cell residual, which contribute for inert biomass. Notice that the active biomass decay rate is much faster than EPS (Laspidou and Rittmann, 2002a). The consumption of electron acceptor in the scheme is added to complete the mass balance. The electron acceptor is used to oxidize the electron donor substrates, the SMP as the recycle products of initial electron donor substrates and the biodegradable part of active inert during endogenous decay. This utilized theory has characterized the formation of EPS or SMP and proposed a complete kinetic model.

### 1.6 Modeling of biological nitrogen removal

The modeling of biological wastewater treatment processes using activated sludge models was officially started and introduced since 1983. The activated sludge models (ASMs) can predict the oxygen uptake rate, COD removal, and nitrogen and phosphorus conversion in wastewater treatment plants. The first ASM, so-called ASM1 model is based on Monod kinetic. The model includes the COD, nitrogen removal, oxygen consumption processes and prediction of the sludge production. The characteristics of wastewater in this model are described in seven dissolved and six particulate components (dissolve oxygen and alkalinity are also included). These components are used to explain the status of two biomass groups and seven fractions of organic matters (COD) and four factions of nitrogen (van Loosdrecht et al., 2015). There are eight processes modeled for the hydrolysis, growth and decay of various living biomass. The model is performed within definition of process rates and their stoichiometric matrix, which contains stoichiometric coefficients and kinetic factors. First line of the matrix represents state variables involved in model processes, and first column displays different processes (van Loosdrecht et al., 2015).

Later, since the enhanced biological phosphorous removal is not cooperated in ASM1, the ASM2 additionally introduced the processes including phosphate accumulating organisms (PAO) in 1995 by (Henze et al., 1995). The approach of ASM1 and ASM2 is similar, when cell is assumed as a black box and metabolic pathways inside the cell are simulated. Later, the model ASM3 was developed in 1999 as a new standard for ASM-based modeling, which proposed an endogenous respiration process instead of decay of heterotrophic to inert biomass and introduction of a storage component for organic compounds (Gujer et al., 1999) (Fig. 12b).

The significant difference of ASM3 and ASM1 is that the ASM3 recognized three rates of oxygen consumption of readily biodegradable COD (rapid rate), slowly biodegradable COD (slow rate) and endogenous oxygen utilization rate (slowest rate). On the other hand, the ASM1 only considers one oxygen consumption process, so it would be problematic for model calibration, because of other processes are indirectly influenced to oxygen consumption process (van Loosdrecht et al., 2015). The second difference of ASM3 with ASM1 is the introduction of storage substrates ($X_{STO, s}$). The test of oxygen utilization rate (OUR) of activated sludge revealed that the living biomass quickly consumes the readily biodegradable COD, then this will be stored as an internal substrate which later can be used for the growth of biomass. This storage term is not considered in ASM1 (Fig. 12a) where a single readily biodegradable substrate e.g. acetate is used as substrate. However, the uncertainty of ASM1 for nitrogen removal systems comparing to ASM3 is very low, because of nitrification is a slow process and the slowly biodegradable COD has sufficient retention time in this model structure.

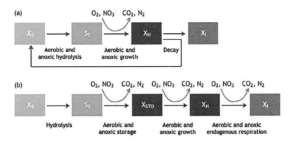

Figure 12: Degradation of organic substrates in a) ASM1 and b) ASM3 (van Loosdrecht et al., 2015)

The first mathematical model of anammox-based reactor for BNR was published in 2004, which included enrichment of anammox employing a sequential batch reactor. The proposed model for the anammox process applied ASM1 extended with two steps nitrification-denitrification (Dapena-Mora et al., 2004). The reactor could successfully reach high and stable nitrogen removal efficiency (>80 %). The model for more than 150 days could predict the effluent data with appropriate level of agreement. The co-existence of heterotrophs in the rector was also observed, which helps to retain anoxic condition by consuming excessed oxygen and prevent the growth of nitrifying bacteria (Dapena-Mora et al., 2004).

### 1.6.1 Modeling of biofilm and granular sludge systems

The modeling of biological wastewater treatment process in aggregated form is essential since biomass tends to aggregate in the biofilms or flocs forms. However, this work is more complex due to existence of substrate gradients, which physical processes are also involved like mass transfer at the bulk/biofilm interface, reaction and diffusion models and detachment (Horn and Lackner, 2014).

The substrates in bulk liquid phase are transported convectively to the biofilm, and a mass transfer boundary layer is developed around the biofilms which diffusion is the main mechanism. This layer is called diffusion bound layer (DBF) with the thickness around 100 μm. The classical film theory has defined mass transfer coefficient ($k_L$) as the ratio of diffusion coefficient ($D_i$) and the thickness of DBF ($L_C$) (Eq. 11)

$$k_L \cong \frac{D_i}{L_C}$$ (Eq. 11)

However, the transport within the boundary layer was measured, which revealed the presence of diffusion and advection. Thus, the classical approach in Eq. 11 is not accurate enough. Sherwood number was introduced, which is the dimensionless number of mass transfer. The

Sherwood number (Sh) is calculated by dividing the total mass transfer flux by the diffusion coefficient of flux ($D_i$) (Eq. 12).

$$Sh \cong \frac{K_L}{L_C} \times d \tag{Eq. 12}$$

where d is length of the reactor geometry. The extension of Sherwood number by considering the enlargement of surface area has proposed, but it has not been yet incorporate into 1D model (Horn and Lackner, 2014).

The reaction of diffusion processes in biofilm depends on substrate concentration. The substrate consumption is based on the second Fick's law of diffusion and a steady state approach. This model was challenged since there are more than one species co-existing in the biofilm (with different growth rate and yield coefficients), the spatial distribution of the biomass along the axis perpendicular to the substratum will be unevenly, results crucially effect to the outcome of the model. In 1986, a publication from Wanner and Gujer has solved this issue and their approach is widely being used in 1D modeling. The living and inert biomass are categorized as particulate matters, and the mass balance of particulate matter in biofilms is describe in Eq. 13 as below (Wanner and Reichert, 1996).

$$\frac{\partial X_i}{\partial t} = -\frac{u_F}{\partial z}\frac{\partial X_i}{\partial z} + r_{X_i} \tag{Eq. 13}$$

where $X_i$ is the concentration of particulate matter; t is the time, $u_F$ is the velocity of particulate matter i moving perpendicular to the substratum; $z$ is value of z axis and $r_{X_i}$ is the rate of conversion of particulate matter $X_i$

The key parameter in the Wanner and Gujer model is $u_F$, which the velocity that can be positive during the growth of biomass, and negative during other processes, such as lysis process. Notice that the biofilm density strongly influences on $u_F$, if this value is small the biofilm thickness will grow faster ($u_F$ increases). For 1D modeling, the biofilm density is relied on experience or measurement rather than prediction.

The detachment occurs when the external shear forces dominates to internal strength of biofilms. The detachment is formulated as the change of thickness of biofilm over the time, and linked with different parameter, such as biofilm density, thickness of biofilm, growth rate, and shear stress. Different approaches of modeling detachment, and how to incorporate with biofilm is proposed in literatures (Horn and Lackner, 2014). The detachment coefficient ($k_d$) is used in all detachment formulas as a lumped parameter. The detachment rate is defined based on the constant substrate load, but also it can be a function of time for example during the back washing.

The biofilm modeling has been widespread developed, while the multi-dimensional models of biofilm enhance virtually the biofilm system and research, the simplified 1D model has applied as most useful engineering tool (Boltz et al., 2017). For 1D biofilm modeling, one of the successful tools applying Wanner and Gujer approach is AQUASIM, which is widely used in 1D model.

### 1.6.2 AQUASIM: a toolbox for biological wastewater treatment modeling

AQUASIM v2.1 is a modeling tool of aquatic system and was developed by EAWAG in Switzerland (http://www.eawag.ch/forchung/siam/software/aquasim/index) (Reichert, 1994). The AQUASIM software allows simulating of multi-substrates and multi- species biofilm systems. The biofilm compartments of AQUASIM are designed into three zones: biofilm solid matrix, bulk liquid, and liquid boundary layer (Fig. 13).

The distribution of living biomass in the biofilm is determined based on the gradient of substrate available along perpendicular to the substratum. The liquid phase is assumed completely mixing and the mass transfer from liquid phase through boundary layer to the biofilm solid matrix is considered as Wanner and Gujer's model. The fluxes are calculated for influent, effluent, transporting across permeable substratum and atmosphere (black arrows in Fig. 13), and exchanging between phases within the biofilm compartment (the grey arrows in Fig. 13). Besides, the attachment and detachment of microbial cells are considered in AQUASIM and defined in the dialog box of biofilm reactor compartment. The properties of the biofilm compartment are specified in the dialog box which allow to define reactor type, the pore volume with or without suspended solids, grid number of biofilm matrix, surface area of the biofilm and the detachment velocity of the biofilm surface (Wanner and Morgenroth, 2004). The variables of substrates or biomass are initialized in the model and the automatic calibration function can calculate their values. The processes occurring in the biofilm compartment are activated in the model, the inactivated processes will be automatically set to zero. This feature is easier for the modeler to apply modification or upgrades to the model. Main applications of AQUASIM are simulation of substrate removal and prediction of particulate components in biofilm solid matrix. Modeling the bulk phase and the microbial interactions in biofilms are based on kinetics of Benchmark Problem 3 (BP3) confirmed by IWA's Biofilm Modeling Task Group (Rittmann et al., 2004). The effluent is calculated as the function of the concentration of the influent and the growth of living biomass in biofilms. Moreover, AQUASIM can predict the attachment and detachment of biofilm, thus the thickness of biofilm is able to be calculated. The microbial relative abundance also can be simulated in the direction perpendicular to the substratum, which has the largest concentration gradient (Reichert, 1998).

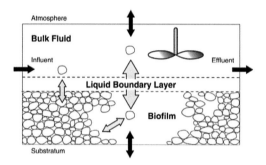

Figure 13: The biofilm compartment of AQUASIM. The black arrows show possible mass transfer across the biofilm compartment, grey arrows show the mass transfer within the compartment (Wanner and Morgenroth, 2004)

The international water association has implemented a variety of biofilm models to determine the best match for typical biofilm problems. Results have shown that the simple analytical 1D model is more practical than complicated two- and three-dimensional model. Since AQUASIM is designed in between simple and complicated modeling systems, it is very useful for complicated biofilm which competition of different microbial groups are significant to be evaluated. At least several knowledge gaps related to biofilm modeling:

i.   Current implications from the mathematical modeling of biofilm reactors focus on simulation of the substrates fate but giving less insight into the active and inactive microbial population with regards to dominance, abundance and biocoenosis. Besides, in terms of real-time dynamic modeling of extracellular polymeric substances (EPS) and soluble microbial products (SMP) in biofilm matrix only few studies are available (Liu et al., 2016; Ni et al., 2012; Xie et al., 2012) especially for EPS and SMP formation of nitrogen-converting bacteria in waste water treatment systems.

ii.  Comprehensive models require such a systematic algorithm for model verification tasks, which has not been adequately addressed in current research articles (Zhu et al., 2016). Besides, due to high range of parameters, the output of biofilm models is prone to numerical instability and changes in kinetic and stoichiometric parameters and sometimes initial parameters. Hence these values must be chosen and identified carefully and in a systematic approach.

iii.    Interoperation of SMP and EPS kinetic to nitrogen removal processes in the models are empirical than mechanistic. There is still a significant knowledge gap and limited research works, especially concerning the values of reaction kinetic parameters for EPS and SMP (BAP and UAP) production and lysis (Seo, 2009). Therefore, to build a more accurate data-driven model, further attentions for kinetic parameterization of EPS and SMP related parameters are required.

iv.    Totally different approaches and models to simulate aerobic and anoxic biofilm reactors for nitrogen removal processes consisting of mass balance, detachment and attachment, biofilm layers' sub-processes and rate equations were presented previously (Kovárová-Kovar and Egli, 1998; Kuai and Verstraete, 1998). Some of these biofilm models have addressed anammox plants (Dapena-Mora et al., 2004; Ni et al., 2009). However, a few works were calibrated and validated with good long-term experimental data of stably operated plants. Some works used valid data for calibration of the models but merely for nitrogen components or organic carbon, but no specific focus was given onto calibration and validation of microbial community dynamics in the biofilm with observed data from molecular approaches.

## 1.7   Motivations, approaches, tools and aim

The rapid establishment of full-scale of anammox plant worldwide requires an enhanced understanding of microbial composition and knowledge about physiological characteristics of anammox species, since anammox bacteria are slow growing microorganism. As biofilm-based and granular technologies are expanding for enhanced nitrogen removal, further investigation through biofilms characteristics and microbial products is vital to optimize the process. In this way, the mathematical modeling is a useful approach for simulation and prediction of the microbial behavior and biofilm characteristic as well as kinetic and physiological evaluations together with laboratory experiments. Also, yet there are limited studies towards the long-term data analysis, feasibility study and stochastic risk and stability assessment of anammox biofilm and granular systems. On the other hand, molecular methods are valuable for studying microbial communities and monitoring the performance of a plant and its mathematical model in complex environments such as anammox reactors. There are many different molecular methods available for use and selecting the appropriate method to be applied depends on the study objectives. The aims can be for example, to (i) identify the microorganisms inside the reactor; (ii) quantify specific groups of microorganisms inside the reactor; (iii) evaluate the microbial community dynamic over time. Therefore, in this thesis three approaches are considered to further investigate and optimize anammox-based processes.: (i) data mining and data analysis approaches including long-term risk assessment for running full-scale plants (ii) mathematical modeling including firstly model-based analysis of short batch experiments and secondly long-term model-based evaluation of a full-scale anammox plant and (iii) lab-scale experimental approaches (Fig. 14). Molecular approaches will be the main backing tool in this thesis for identification, (semi)quantification and dynamical evaluation together with basic physiochemical evaluation e.g. (pH. T, COD, nitrogen components, etc.) of wastewater contents and empirical evaluations such as calculation of parameters e.g. (MLSS, MLVSS, SRT, SVI, etc.). Hence the main aim of this research was to study the biological feasibility of various anammox-based technologies to remove nitrogen from different wastewaters and combine it with surveying, modeling, and experimental approaches.

## 1.8    Systematic milestone plan of the research

As explained in Fig. 14, in the first milestone (Chapter 2), the survey-based data analysis methodology was applied to determine the performance of biological treatment for high ammonium removal from landfill leachate. The plant was upgraded combining the activated sludge process followed by activated carbon reactor and it is running stably for more than a decade since end of 2001. Based on a long-term analysis of data collected from 2006 to 2015, the average total nitrogen removal efficiency achieved for wastewaters with various C: N ratios varying will be studied and system efficiency and performance as well as stochastic failure risk of the plant with and without the presence of activated carbon biofilm reactor will be evaluated.

In the second milestone (Chapter 3), a short-term mathematical modeling approach was used based on anoxic batch experiments in the lab to get idea through kinetics and physiology of dominant bacterial groups and estimate the unknown kinetic and physiological parameters of anammox species. In this case simultaneous anammox and denitrification (SAD) process was simulated. The proposed parameter identifiability and estimation technique in this chapter is a useful tool to improve the model performance.

In the third milestone (Chapter 4), long-term modeling approach for biological nitrogen removal in an anammox-rich granular sludge reactor was examined. The main aim of the work is to develop and compare two with and without SMP and EPS definitions to explain the abundance of independent microbial groups in terms of relative bio-volume fraction, to calibrate and validate the model with chemical and microbiological data and to and to further describe bacteria growth through biofilm layers.

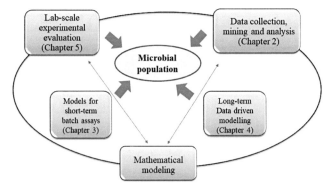

Figure 14: Scheme of systematic plan of research and the applied research approaches (grey boxes) and the main backing tool (yellow) based on multidimensional approaches applied in this thesis to investigate and optimize anammox-based systems: data mining, modeling (short-term and long-term data driven verified models) and lab-scale experimental evaluation of a biofilm reactor.

In the fourth and last milestone of this research (Chapter 5), design and experimental investigation of two reactors with fixed-film technologies for biological nitrogen removal was aimed. Other objectives of this chapter are (i) to operate two labs-scale SBRs one as IFAS and another with combination of biofilm on activated carbon and activated sludge flocs and to compare which reactor has a better potential to reach a faster start-up and a higher efficiency during short time; (ii) to analyze the microbial community diversity in both systems; (iii) to evaluate effect of reduced temperature gradient on nitrogen removal performance and specific activities; and (iv) to analyze the effectiveness of various aeration patterns under reduced temperature on nitrogen removal performance and microbial community structure. In all parts, fluorescent *in situ* hybridization (FISH) and quantitative

FISH (QFISH) were the main backing tool for microbial diversity analysis on floccular sludge and granular samples. In the fourth part, Next-generation Sequencing (NGS) of 16S rRNA gene amplicons was another backing tool for microbial evaluations (Fig. 15).

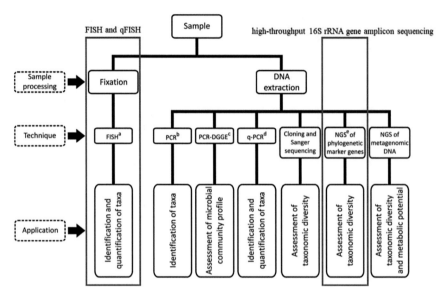

Figure 15: Most commonly used methods to study microbial communities in anammox reactors and their potential applications. (a) Fluorescent in situ hybridization (FISH); (b) polymerase chain reaction (PCR); (c) PCR followed by denaturing gradient gel electrophoresis (DGGE); (d) qPCR; (e) next generation sequencing (NGS). (The graph is adapted from Preira et al., 2017)

## 2 More than a decade of experience of landfill leachate treatment with a full-scale anammox plant combining activated sludge and activated carbon biofilm

### 2.1 Introduction

For an effective treatment of wastewaters with high ammonia concentration such as landfill leachate, activated sludge with nitrification and denitrification can be a primary biological treatment solution. However, the application of the conventional process was limited for wastewaters with high nitrogen contents due to the high operational cost since the introduction of two energy saving and shortcut pathways for nitrogen elimination including: simultaneous nitrification and denitrification (SND) via nitrite or nitrite shunt and partial nitritation-anaerobic ammonium oxidation (anammox) or deammonification (Regmi et al., 2013). As explained, deammonification is an economic autotrophic process conducted by ammonium oxidizing bacteria (AOB) and anammox planctomycetes with repression of nitrite oxidizing bacteria (NOB) (Siegrist et al., 2008). If denitrification process is simultaneously observed in the reactor, the terms simultaneous partial nitrification, anammox and denitrification (SNAD) (Daverey et al., 2015) or simultaneous anammox and denitrification (SAD) can be used (Li et al., 2016). Lackner et al., (2014) surveyed that in terms of reactor configuration the sequencing batch reactors (SBRs) and in terms of biomass type granular systems and moving bed biofilm reactors (MBBRs) are mostly applied. Better anammox activity and preservation can be achieved in granular sludge either formed freely (Vlaeminck et al., 2008; Winkler et al., 2012), granular sludge formed on granular activated carbon (GAC) (Wenjie et al., 2015), biofilms fixed on immobile carriers (Zhang et al., 2015) as well as biofilms formed on suspended artificial carriers (Christensson et al., 2013),. However, there are limited studies towards the long-term data analysis and risk assessment of anammox biofilm systems.

In terms of ecophysiology, different anammox bacteria have a niche specificity and the six known genera of anammox bacteria are divided into related lineages phylogenetically as: *Candidatus* Scalindua, *Ca.* Kenuenia, *Ca.* Jettenia, *Ca.*Annamoxglobus, *Ca.* Anammoxim icrobium and *Ca.* Brocadia (Kartal et al., 2007; Kuenen, 2008; Khramenkov et al., 2013). Previous studies remarked privileged physiological characteristics for *Ca.* Scalindua compared to other anammox species (Awata et al., 2013). Other reports highlighted the existence of the genus *Ca.*Scalindua mainly in natural habitats such as oxygen minimum zones in soils and marine environments and only few reports found this genus in engineered systems either in a lab-scale (Tsushima et al., 2007) or a pilot scale plant (Schmid et al.,

2003). Yet, no study could prove the long-term, stable and remarkable existence of *Ca.* Scalindua species in forms of flocs and biofilms in an anammox full-scale plant. This work combining research and application survey methods, aims to evaluate long-term monitored data of a combined activated sludge- activated carbon biofilm system to prove how this technology works as a robust and novel tool to enrich anammox biomass in two forms of suspended flocs and granular biofilms. The specific objectives are: (1) evaluating the consistent and stable treatment of total nitrogen for various COD/TN ratios; (2) developing a risk assessment approach using long-term actual data; (3) detecting the bacterial community in the system using a quantifiable molecular cytogenetics method with specific focus on the existence and relative abundance of *Ca.* Scalindua sp. and (4) calculating the reductions in ongoing costs related to excess sludge management, organic carbon and energy consumption.

## 2.2 Materials and methods

### 2.2.1 Operation of the plant

The landfill leachate treatment plant is in Zentraldeponie Emscherbruch (ZDE) which is a municipal waste disposal site (Herten, Germany). The plant was primarily operated as a conventional activated sludge plant with cascade nitrification and denitrification. That caused more external organic carbon source, higher surplus sludge and additional aeration. The plant was renovated in 2001 by adding two further processes of ultrafiltration and granular activated carbon after activated sludge and before the outflow. The main inflow rate is $30 \text{ m}^3 \text{ h}^{-1}$ or $260,000 \text{ m}^{} \text{ year}^{-1}$. The influent stems from solid waste leachate via screw pumps with an initial storage of six buffer tanks with individual volume of 900 to $1,300 \text{ m}^3$. The first step consists of the activated sludge process incorporating one denitrification basin and three nitrification basins with a volumetric capacity of $165 \text{ m}^3$ per basin. The second part of the plant consists of four parallel ultrafiltration membranes (UFs) lined by $0.8 \text{ μm}$ porous filters with a total surface area of $300$–$350 \text{ m}^2$. The last treatment before the outflow consists of activated carbon reactors where the physical process of COD reduction by adsorption arises. Moreover, activated carbon reactors are associated with biological treatment where GACs serve as biofilm carrier for attachment of microorganisms and increasing the biomass retention.

## 2.2.2 Sample preparation

From December 2015 to October 2016, collection of two types of samples from the activated sludge and activated carbon biofilm reactors was performed monthly. Aliquots of 5 ml of samples were collected and diluted with deionized water to reach 35 ml. Granular samples were mixed, dispersed and sonicated in an ultrasound for 15 min to loosen the biofilm matrix. For flocs of activated sludge, a mechanical pre-treatment procedure by glass beads were applied. Then, samples were centrifuged at 4,000 rpm for 5 min and washed with deionized water. For further cytogenetics purposes, cells were fixed with 4 % (v/v) paraformaldehyde in PBS for 1 h at room temperature, followed by two washing steps with PBS followed by centrifugation at 10,000 rpm for 10 min. Fixed cells were suspended in a 50 % (v/v) PBS/ethanol solution and stored at $-20$ °C.

## 2.2.3 Fluorescent in situ hybridization (FISH)

To check the dominancy of microbial groups, for all monthly collected samples, FISH adapted from the protocol applied by Azari et al. (2016) using six group-specific 16S rRNA-targeted oligonucleotide probes was performed. Specific probes were labelled at 5′ position with the fluorochrome Cy3. Anammox specific probes S-*-Amx-0368-a-A-18 for all anammox planctomycetes, Amx- 0820-a-A-22 for Ca. Brocadia and Ca. Kuenenia-like anammox bacteria, Sca-1309-a-A-21 for the genus Ca. Scalindua and Nso190 probe for betaproteobacteria ammonia-oxidizing bacteria (AOB) were used. To target NOB, NIT3 for Nitrobacter sp. and Ntspa662 for Nitrospira sp. have been applied. For the visualization of most of bacteria, FISH using EUB 338 mix probe covering 90 % of domain bacteria with fluorochrome Cy5 was used (Biomers.netGmbH, Ulm, Germany). Additionally, to check the hybridization efficiency, microscope slide can be covered with 4′,6-diamidino-2-phenylindole (DAPI) solution with concentration of 1 μg ml$^{-1}$.

For all samples, image acquisition was done on Zeiss AxioImager M2 epifluorescence microscope. Besides, Zeiss LSM 510 confocal laser scanning microscope (CLSM) was used to observe more precise situations (Carl Zeiss, Jena, Germany).

To estimate the abundancy of each microbial group, additional steps for quantitative fluorescent in situ hybridization (qFISH) analyses were performed for nine samples. For each sample, at least twenty random fields of view (FOV) were imaged to determine the average relative abundance of each microbial group. Quantification of average relative abundances (so called biovolume fractions) was done by calculating the area taken up by fluorescence targeted cells from specific probes compared to the area complimentary to the EUB 338 mix probe. This step was performed using daime software (version 2.1) which is an image processing and analysis tool for semi-quantification of complex microbial systems by splitting the images into individual color channels followed by image segmentation (Daims and Wagner, 2007). The procedure for qFISH is explained in the appendix A.

### 2.2.4 Measurements and statistical data analysis

The parameters evaluated in influents, effluent from activated sludge reactors and the final effluent were: pH, the electrical conductivity, chemical oxygen demand (COD) and nitrogen components. Nitrogen analyses included daily measurements of ammonium as $NH_4^+$-N, nitrite as $NO_2^-$-N, nitrate as $NO_3^-$-N and COD based on German standards using Hach Lange cuvette test kits and Hach Lange spectrophotometer (DR2800, Hach, USA). The total nitrogen (TN) per mgN $L^{-1}$ in the effluent was estimated as sum of $NH_4^+$-N, $NO_3^-$-N and $NO_2^-$-N expressed as mgN $L^{-1}$. The amount of organic nitrogen present in the filtrate wastewater was ignorable. For the influent only $NH_4^+$-N was recorded since $NO_2^-$-N and $NO_3^-$-N were negligible. All daily measurements for TN and COD were averaged on monthly basis. Thereafter the nitrogen removal efficiency (NRE) was determined for the daily and monthly values based on Eq. 14 according to Niu et al. (2016):

$$NRE = 100 \times \frac{TN_i - TN_o}{TN_i} \qquad \text{(Eq. 14)}$$

$TN_i$ in mgN $L^{-1}$ is the total nitrogen concentration in the influent and $TN_o$ in mgN $L^{-1}$ is the total nitrogen concentration in the effluent. If the effluent is the outflow after denitrification and nitrification by activated sludge stage, the result is called the NRE for the treatment merely by activated sludge. But if the effluent is the final outflow of the plant after treatment by biofilms on GACs, the result is called final NRE for combined treatment. The difference can elucidate that how much nitrogen was removed by biofilms grown on GACs.

An analysis of risk assessment based on long-term real data was developed as a deterministic model describing the concentrations and treatment efficiencies and the level of exceedances. The risk interpretation was formulated in the form of concentration – duration – frequency (CDF) curves based on exceedance probability analyses generated by dividing the norm exceedance times into several classes according to Rousseau et al. (2001).

## 2.3 Results and discussion

### 2.3.1 The morphology and structure of the granular biofilm

The morphology and granular size of the mature anammox biofilms are shown in Fig. 16. GACs act as external physical carriers for developing anammox based granular biofilms. Sludge color is an indicator of sludge viability and biological performance. The bright red color represents for small and fresh anammox granules (Fig. 16a, samples taken on February 2016). The carmine red, brown red and brick red colors of settled sludge was more representative for matured anammox granule (Fig. 16b, samples taken on April 2016). The mature anammox granule was characterized by a cauliflower-like shape (Fig. 16b) and the fresh and immature granules had mostly spherical and elliptical shapes (Fig. 16a). This fact is in accordance with other findings by Tang et al. (2011), Lu et al. (2012) and Ke et al. (2015). Layers of aggregated biomass were separated from each other by interstitial spaces and voids (Fig .16c). Interstitial voids were surrounded by extracellular polymeric substances (EPS) and were filled with water or dinitrogen gas. The size of interstitial spaces is proportional to the size of granule and voids could serve as water channel and dinitrogen gas storage which is the product of the anammox process (Ali et al., 2013). It is also explained by Weissbrodt et al. (2016) that if a granular biofilm sludge is more aerobic, the biofilm forms more mobile biofilms with a gel-like consistence and more water phase inside the interstitial spaces. To describe the reasons for the stability of activated carbon-activated sludge system, over more than a decade, EPS formation plays a key role. The microorganisms in biofilms live in a self-produced matrix of hydrated EPS that form their immediate environment. Due to findings by Flemming and Wingender (2010), EPS matrix makes biofilms as a stable state of life on earth. In terms of size distribution, collected mature and fresh biofilms of granules were irregular in shape with a diameter varying between ~0.8–24 mm (Fig. 16d). Of interest was also the identification of shape of planctomycetes-like bacteria. Therefore, the biofilm samples were diluted, stained with DAPI and checked under epifluorescence microscopy. The typical appearance of the relatively large anammox planctomycetes is determined by the presence of a large internal compartment, the anammoxosome, responsible for the anammox metabolic reactions (Fig. 16e and f). Depending on the position of the anammoxosome, the outline of bacteria may appear in circular, cocci, sickle or crescent shape under the microscope, because ribosomal RNA, which is the target of FISH probes, only exists outside of the vacuole inside the riboplasm. Furthermore, anammox bacteria are protected by a complex membrane and cell wall system, having an outer cell wall and two inner cell membranes (van Niftrik et al., 2004).

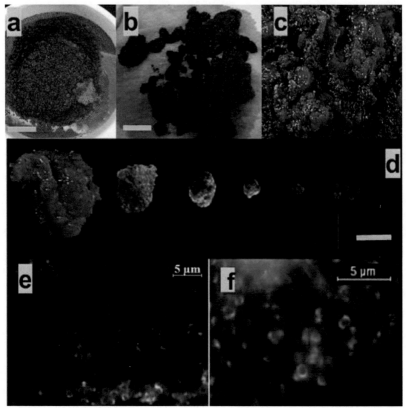

Figure 16: (a) Fresh granules from samples taken in February 2016 (b) red and brown mature granules and (c) inner section of granules for samples taken in December 2015, (d) the morphology and physical structure of granules distributed in different sizes, (e and f) magnified cells of planctomycetes-like bacteria from diluted and pre-treated granular samples directly stained with DAPI and directly applied for epifluorescence microscopy. The sickle, crescent, ring or cocci shaped outlines of bacteria indicates big vacuoles which is typically found in anammox planctomycetes. Samples were taken from activated carbon biofilm reactors. (bar 1 cm).

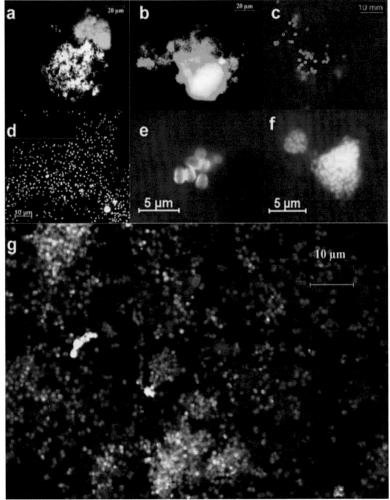

Figure 17: FISH micrographs with: AMX820 specific probe in Cy3 dye (orange) for granules from biofilm reactor (a and g) and flocs from activated sludge reactors (d) hybridized with 10 % formamide to target species *Ca*. Brocadia anammoxidans and *Ca*. Kuenenia stuttgartiensis, Sca1309 specific probe in Cy3 dye (orange) for granules from biofilm reactor (b and c) and flocs from activated sludge reactors (e) hybridized with 5 % formamide to target genus *Ca*. Scalindua and Nso190 specific probe in Cy3 dye (orange) for flocs (f) hybridized with 45 % formamide to target to target βAOB. EUB 338 probe in FITC dye (green) was used to target most of domain bacteria. Combined signals will be in yellow color. image acquisitions were done by epifluorescence microscopy except image (g) with CLSM.

**2.3.2    Fluorescent in situ hybridization and the existence of *Ca.* Scalindua genus**

During the entire sampling period, anammox bacteria remained significant in abundance for flocs in activated sludge (Fig. 17d,e and Fig. B1,2) and for biofilms in granules (Fig. 2a,b,c,g). Fig. 17f also showed the identification of nitrifiers in the activated sludge. In general, the results complied that the anammox population of the sample of this plant are strongly represented while AOB and NOB population was less than anammox. The results agree with other similar studies (Kindaichi et al., 2007; McSwain et al., 2005). *Ca.* Scalindua was found only in limited single cells in flocs (Fig. 2e and Fig. S3). But anammox bacterial population includes higher concentration of *Ca.* Scalindua sp. in granular biofilms forming both colonies and single cells (Fig. 17b and c). FISH analysis confirmed the formation of the species *Ca.* Scalindua and other anammox-like bacteria in the middle of the inoculated biofilm sample (Fig. 17a and b). This compromises with previous reports claiming anammox bacteria can dominate the granule interiors and maintain its activity in oxygen limiting conditions (Kindaichi et al., 2016).

In terms of abundancy using qFISH, 9 samples over time were collected and the average relative abundance for activated sludge samples was of 32 - 57.3 % for genera *Ca.* Brocadia and *Ca.*Kuenenia (AMX820), of 15.2–17 % for the genus *Ca.* Scalindua and of 8.9–18.2 % for AOB (Nso190). For granular biofilms, relative abundance was of 47–64.9 % for genera *Ca.* Brocadia and Kuenenia (AMX820), of 15.9 -26.8 % for the genus *Ca.* Scalindua, of 4.6–5.4 % for AOB (Nso190) and of less than 1 -3.3 % for NOB (Ntspa662 and NIT3).

The occurrence of *Ca.* Scalindua genus in ZDE plant is not only backed up by FISH. A close matching to *Ca.* Scalindua marina species was primarily reported by Ke et al. (2015) in a similar plant using the phylogenetic analysis of AnAOB based on 16S gene followed by DNA extraction and PCR amplification. Therefore, the authors set up further FISH experiments using *Ca.* Scalindua specific probe (Sca1309) to prove the accumulation of this genera in this plant. Understanding the reasons of this accumulation is crucial since almost all relevant literature in the wastewater context reports almost other genera but *Ca.* Scalindua. Schmid et al. (2003) found this genus by FISH in one small pilot unit installed in a leachate treatment plant with the capacity of 8.5 m$^3$ h$^{-1}$ and Tsushima et al. (2007) found similar genus in a lab-scale up-flow fixed-bed glass biofilm column reactor using DNA sequencing. But in this work, we remarkably found *Ca.* Scalindua in multiple samples in two forms of flocs and biofilms in different time periods from the anammox full-scale plant polishing more than 30 m³ h⁻¹wastewater. The existence of *Ca.* Scalindua in this anammox plant is of high importance since previous study (Awata et al., 2013) showed privileged physiological characteristics compared to other anammox species. The specie is known to be capable of tolerance to severe conditions because of its potential origin from marine ecosystem. temperatures of 10–25 °C, pH of 6–8 and high salinity of 1.5–4 mmol which is important for treating ammonium-rich wastewaters such as landfill leachate that contains high salt concentrations. It also has higher affinity for nitrite and a lower growth rate and biomass yield. The higher affinity for nitrite is necessary for the bacteria to survive in an environment with extremely low levels of nitrite and ammonium concentrations.

Understanding the source of aggregation of this marine genus is also crucial because anammox genetic differences have fundamental variations in their physiological characteristics. A biogeographical distribution pattern was reported over a large area from terrestrial ecosystems, associated with pollution and remarkable anthropogenic impact, particularly non-pristine terrestrial ecosystems (Han and Gu, 2015; Li et al., 2013). In this pattern, *Ca.* Scalindua genus was discovered widely in oceans (Hong et al., 2011; Han and Gu, 2015) and freshwater wetland and agricultural land (Lee et al., 2014, 2016; Wang and Gu, 2013). Our current results show that *Ca.* Scalindua genus can also be notable in wastewater system, enabling a high possibility of divergence of this genus from the native one detected primarily in oceans (Hong et al., 2011) or from a non-contaminated wetland (Wang and Gu, 2013; Lee et al., 2016). Recently, the new species of *Ca.* Scalindua have also been reported in the South China Sea, including subclusters of *Ca.* Scalindua zhenghei-I, zhenghei-II, and zhenghei-III (Hong et al., 2011) and coastal mangrove wetland (Han and Gu, 2015). These new species show different dynamics in coastal water and the open ocean, suggesting specificity of selective species to niche preference (Hong et al., 2011; Han and Gu, 2015).

### 2.3.3 Long term nitrogen removal performance analysis

Following upgrading of the plant in 2001, the microbial community shifted towards granulation of red biofilms containing anammox bacteria. Consistent and flexible nitrogen removal was monitored for more than fifteen years with a maximum daily nitrogen removal efficiency of 98.3 % and a removal rate of 1.1 kgN m⁻³ d⁻¹. The plant has been operating continuously and stably between temperatures fluctuating from 16.8 to 34.7 °C, without any serious failure and the total nitrogen and COD removal efficiency for combined treatment was monitored above 90 % (Table B1 and Fig. B4). The electrical conductivity in leachate samples over the period of data analysis, ranged from 1.1 to 25.2 mS cm$^{-1}$ in the influent and from 4.4 to 13.7 mS cm$^{-1}$ in the final effluent. The pH was from 5.8 to 7.9 in the effluent. Initial results of non-continuous data analysis from 2002 to 2005 demonstrated that the operation of ZDE plant might be the earliest anammox full-scale plant in the world. Our data dispute the earlier statement by Van der Star et al. (2007) claiming that the first full-scale anammox reactor into the operation is the anammox treatment plant in Netherlands.

Previous methods such as real-time polymerase chain reaction (qPCR and activity batch tests on the biomass previously revealed the coexistence of anammox bacteria, denitrifiers and AOB in the activated sludge in the form of flocs and biofilms with dominancy of anammox bacteria after upgrading the plant (Rekers et al., 2007; Ke, 2014; Ke et al., 2015).

Due to the continuous recorded data for 2007–2015, the monthly value of final NRE for combined treatment ranged from 84 % to 97.2 %. The average of the NRE value for combined treatment was 94 % ± 2.7 % in monthly basis. The minimum efficiency of the activated sludge system was observed during February 2009 and the maximum efficiencies were achieved in September 2009. During this period the NRE for treatment by activated sludge process was only 78 % and 19.4 % of total nitrogen in influent was treated by biofilms attached to GACs. The results also elucidate that by average, up to more than 20 % of the nitrogen in the influent can be eliminated and compensated by embedded bacteria to GACs as a supplementary process after activated sludge. Monthly average values for the nitrogen loading rate (NLR) over the entire studied period, were 0.71 kgN m$^{-3}$d$^{-1}$ ± 0.1 kgN m$^{-3}$ d$^{-1}$ and ranged from 0.45 kgN m$^{-3}$ d$^{-1}$ to 1 kgN m$^{-3}$ d$^{-1}$ (Fig. 18a). From this amount, 0.61 kgN m$^{-3}$ d$^{-1}$ ± 0.12 kgN m$^{-3}$ d$^{-1}$ corresponding to 83 % of nitrogen available in the influent was eliminated by activated sludge process (Fig. 18a, black and white patterns) and 0.07 kgN m$^{-3}$ d-$^{1}$ ± 0.1 kgN m$^{-3}$ d-$^{1}$ was removed by microorganisms attached to the activated carbon forming a biofilm (see Fig. 18a black parts). The results also highlight that in some months, the activated sludge process eliminated only 60 % of the total nitrogen in the effluent. Finally, an average of 0.67 kgN m$^{-3}$ d$^{-1}$ ± 0.11 kgN m$^{-3}$ d$^{-1}$ corresponding to 94 % of the total nitrogen available in the influent was eliminated by a combined activated sludge-activated carbon biofilm process. Only the average total nitrogen of 0.04 ± 0.01 kgN m$^{-3}$ d$^{-1}$ remained untreated in the effluent (Fig. 18a, grey parts). The maximum treatment efficiency was reached when the untreated total nitrogen in the effluent was 0.02 kgN

$m^{-3} d^{-1}$ in September 2009. During this period, the NRE for combined treatment was 97.2 % while the NRE for treatment by activated sludge process was only 87 %. On the other hand, in February 2011 when a total maximum amount of nitrogen in the effluent equal to 0.09, remained untreated, the NRE for combined treatment was 88 % which is still acceptable and corresponds to a high efficiency in the plant. Similar hybrid and combined case studies with high nitrogen load such as Reeve et al. (2016) and Kosari et al. (2014) showed the maximum average TN removal efficiency including anammox, nitrification and denitrification varies between 80 % and 87 % which is lower than ZDE plant. Furthermore, Zhang et al. (2015) addressed an integrated fixed-biofilm system for an activated sludge reactor to enrich anammox and nitrifiers in one stage pilot-scale reactor while NRE was only 80 % for 100 days of operation and nitrogen removal rate (NRR) ranged between 0.4 and 1.1 kgN $m^{-3} d^{-1}$. Some other studies such as Takekawa et al. (2014) showed instability and the deterioration in the performance of the reactor. These studies could achieve neither a high-performance nitrogen removal like ZDE plant during full scale operation nor a long-term stableness and capability of granular formation. Fig. 18b represents the variations over the period 2007 to 2015 for average monthly values of the total nitrogen available in the influent and effluent after treatment by activated sludge and after the combined treatment. The average of total nitrogen (TN) for the influent, effluent after activated sludge processes and after combined treatment was 711 ± 116, 106 ± 30 and 46 ± 13 mgN $L^{-1}$ respectively. The minimum monthly average concentration of TN was observed at 22 mgN $L^{-1}$ while the maximum value was 91 mgN $L^{-1}$. However, the nitrogen removal process seemed stable since the average the total nitrogen concentrate was always less than 50 mg $L^{-1}$. Fig. 18b suggests that elevated levels of ammonium will not influence on lowering NRE and disrupting the total efficiency.

### 2.3.4 Risk assessment

A risk assessment tool for evaluating wastewater treatment performance based on the long-term data of nitrogen removal efficiency of the plant from 2006 to 2015, including and excluding the activated carbon biofilm reactor was analyzed and presented (Fig. 19a). According to the EU Directive concerning urban wastewater treatment(91/271/EEC), the TN in the outflow of a sewage treatment plan must be at least 70–80 % lower than the concentration in the plant inflow. In the design of activated sludge systems, a conventional approach in dealing with uncertainty is implicitly translating it into above-normal safety factors, which in some cases may even increase the capital investments by an order of magnitude. To obviate this problem, the cumulative distribution function (CDF) of random-variables NRE i.e. x was evaluated, bearing the probability in mind that NRE will take a value less than or equal to x. This was defined as a deterministic model for ZDE plant. In Fig. 19a the horizontal axis indicates the level of expected nitrogen removal efficiency and the vertical axis shows the cumulative NRE distribution. The variability due to time is captured in the cumulative distribution curve. To explain the meaning of these figures, the long-term actual results of the WWTP over the years 2002–2015 have been used. During this period, more than 82 % of the TN measurements in the effluents were over the limit of 90 % removal efficiency. Therefore, the plant post modification achieved to an efficiency higher than the EU criteria by achieving to a total NRE of more than 84 % in all studied data points. The plant has never failed the standards although without considering the activated carbon process the likelihood that the total nitrogen removal efficiency is lower than defined criteria of 80 % is more than 24 %. But in case of using the combined system, the likelihood that NRE drops lower than 80 % is very close to zero. The possibility to reach NRE between 80 % and 90 % for the combined system is 25 % and the possibility to that NRE exceeds more than 90 % is 75 %. This fact has been clearly illustrated in Fig. 19a where the cumulative distribution function for two different approaches has been shown, the first graph only with the activated sludge system and the second graph with the combined system.

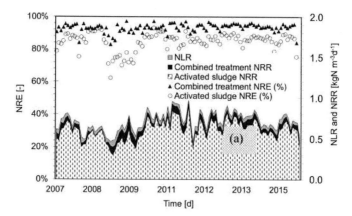

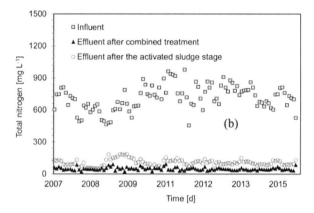

Figure 18: (a) NRE for combined treatment and treatment only after activated sludge stage, NLR and NRR for only biological treatment by activated sludge and combined biological treatment of activated sludge with activated carbon process. (b) Total nitrogen concentration for influent, effluent after treatment with activated sludge and the effluent after combined treatment by activated sludge and activated carbon process. All values were averaged in monthly basis using daily recorded data.

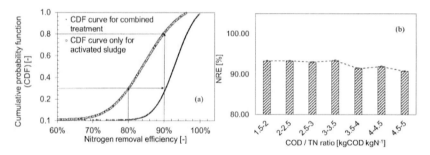

Figure 19: (a) CDF curves for different TN removal efficiency limits for combined treatment and only activated sludge based on the TN concentration for the effluent and influent for time series averaged on monthly basis. NRE limit of exceedances were divided in 12 classes per year. (b) The effect of change in average COD/TN ratio on NRE.

The long-term risk assessment tool can be helpful for probability-based design and it enables engineers to utilize and incorporate it into their design of WWTPs using the available and future information depending on the effluent standards and the climate conditions (Talebizadeh et al., 2014). Fig. 19a also can examine the associated risk before and after the upgrade of the conventional activated sludge plant towards a combined activated sludge and activated carbon system complying stricter effluent standards on nutrients discharge.

### 2.3.5  Effect of COD/TN ratio on nitrogen removal efficiency

According to Fig. 19b during the time frame 2006–2015, the daily values for carbon to nitrogen COD/TN ratio ranged between 1 and 5 kgCOD kg$^{-1}$-TN. By increasing in COD/TN ratio as expected, there is a slight reduction in TN removal efficiency, but this effect is ignorable, and it never affected the stability and long-term efficiency of the process. One column graph with standard deviation bars and the trend line will be provided. The plant is significantly consistent over various COD/TN ratio and stable nitrogen and carbon removal efficiency of more than 90 % was achieved in the plant for various COD/TN contents.

### 2.3.6  Reduction in excess sludge, organic matter and energy

By advancing the plant, the energy efficiency assessment showed a reduction in the specific energy demand per volume of treated wastewater from 1.81 kWh m$^{-3}$ in the beginning of 2002 to 0.135 ± 0.37 kWh m$^{-3}$ for 2005–2015. This change is plausible due to the aeration energy reduction as the main source of energy demand. It shows that by upgrading the conventional nitrification-denitrification activated sludge plant to the anammox plant using combined activated sludge-activated carbon approach the aeration and consequently the energy demand of the plant was reduced up to 87 %. Since the specific energy demand dropped 1.675 kWh m$^{-3}$ and considering that the flowrate was estimated as 260,000 m$^3$ year$^{-1}$, the annual saved energy after renovation is 435,500 kWh year$^{-1}$. The specific energy demand is less than 0.1 kWh m$^{-3}$. (Fig. 20a). In another survey the power consumption of compressors as main unit of energy for nitrification has been analysed from 2002 to 2005. During this time the energy demand for compressors reduced from 850,000 kWh to less than 700,000 kWh month$^{-1}$. The time of the operation of the compressors declined from 1,600 h per month to 600 h per month. Furthermore, the air flowrate for nitrification has been reduced from 1,000 m$^3$ h$^{-1}$ to 200 m$^3$ h$^{-1}$.

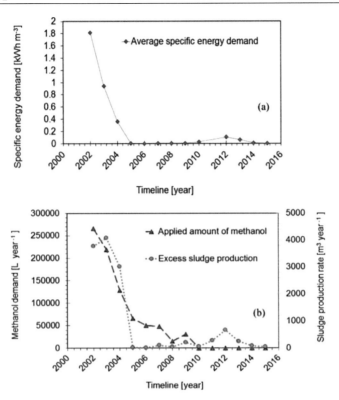

Figure 20: Evaluation of specific energy demand reduction per volume of treated wastewater (a) and reduction in the excess sludge production and organic matter supply (b) during the year 2002–2005 (transition timeframe during upgrade of the plant) and 2005–2015

The average organic carbon consumption was reduced by 91 % while the average surplus sludge production was lowered by 97 % (Fig. 20b). The applied amount of methanol has been reduced from 265,600 L year $^{-1}$ to no methanol addition which occurred during 2010–2016. Moreover, excess sludge production reduced from 3,780 m$^3$ year $^{-1}$ in the beginning of 2002 to 43 m$^3$ year $^1$ at the end of 2015. The total costs including the energy dropped more than 230,000 Euro from 2002 to 2005. Considering 60 Euros per m$^3$ for excess sludge, more than 224,400 Euros per year can be saved due to reduction in excess sludge production. The saved costs due to the addition of the organic carbon source was estimated at 156,100 euros per year in the beginning of 2002 compared to the period 2010 to 2015. The additional costs for methanol addition are compensated. The required monthly amount of carbon in terms of applied methanol to treat one unit of loaded total nitrogen in the influent expressed as kg-C $kg^{-1}$-N was measured (Fig. 20b). After adjustment in the plant, this value was reduced from 1.16 to 0.42 kgC $kg^{-1}$N which shows more than 64 % reduction, and this can potentially save up to 380,000 euros per year only for annual operational costs.

## 2.4 Conclusion and remarks

The performance of a biological treatment plant treating high-ammonium-strength landfill leachate in 2006–2015 was studied. The plant was upgraded combining activated sludge with activated carbon biofilm. The effect of C:N ratio on consistent and stable anammox was discussed, the long-term risk assessment was conducted, and the bacteria community was analysed. The genus *Ca.* Scalindua which has been never reported considerably in other full-scale plants, was detected in flocs and biofilms attached to activated carbon. Excluding the activated carbon biofilms contribution, the average NRE was 23 % less. Besides, adjustments in the plant was proved to be a sustainable tool for integrated waste management leading to a significant reduction in energy, methanol consumption and excess sludge costs.

# 3 Simulation of simultaneous anammox and denitrification for kinetic and physiological characterization of microbial community in a granular biofilm system

## 3.1 Introduction

Combination of partial nitritation and anammox process (PN/A) is considered as a great alternative for treating ammonia-rich wastewater. However, anammox process produce nitrate in the reaction, which leads to a nitrogen reload into the wastewater and restrict the total nitrogen removal rate to maximal 90 %. For stoichiometry of anammox process, most researchers followed Eq. 6 (Strous et al., 1998). Later the anammox process was slightly revisited (Eq. 7) claiming that the production of nitrate as end-product during the anammox process is 38 % less than what it was given before.

Simultaneous partial nitrification, anaerobic ammonium oxidation and denitrification (SNAD) in oxygen-limited conditions (Azari et al., 2017a; Wang et al., 2010) (in case of applying one stage PN/A) or simultaneous anaerobic ammonium oxidation and denitrification (SAD) in anoxic/anaerobic conditions (Li et al., 2016) (in case of applying two-stage PN/A) is a plausible solution for this flaw. Concerning enrichment of the anammox and denitrifying organisms, biofilm systems seem to be especially suited in which the necessary sludge age of more than 20 days can be ensured. A better bacterial activity and higher solids retention time is achievable in granular sludge and in biofilms formed on membranes, plastic carriers, foams or granular activated carbon (Azari et al., 2017a; Vlaeminck et al., 2008). During recent years, mathematical biofilm models for multi-species anammox-based plants proved to be substantial for cost reduction (Ostace et al., 2011), optimizing the system performance and operating conditions (Ni et al., 2009) and prediction of greenhouse gases emission (Ni et al., 2013). With respect to the physiological and kinetic parametrization, different numerical modeling approaches and experimental setups have been investigated (Dapena-Mora et al., 2004; Hao et al., 2002; Ni et al., 2014; Vangsgaard et al., 2013a). However, mathematical complexity is a drawback for the development and application of such biofilm models. Biofilm mathematical models with less variables and parameters can speed up the calculation time. Simplification of such biofilm models is often formulated by using a less complicated form of mathematical expressions (Boltz et al., 2010; Perez et al., 2005), simplifying or combining the governing processes (Smitshuijzen et al., 2016) or using empirical approaches by defining two or more bioreactors in series, to roughly mimic the hydrodynamics of a complex waste water treatment plant (Perez et al., 2014). Yet simplified models mainly have already been applied to examine the interaction anammox bacteria and heterotrophic bacteria and a variety of structures have been formulated such as biofilm models with the definition of soluble microbial products and advection diffusion-based models (Cema et al., 2012; Liu et al., 2016b; Ni et al., 2012, 2009; Vangsgaard et al., 2013b).

At least three drawbacks in current model-based evaluation of anammox-based biofilm systems exist. First, updated stoichiometric parameters based on Eq. 7 are not used and mathematical models still applied old stoichiometry based on Eq. 6. Second, current models representing the anammox process use similar kinetics and physiological parameters for all group of species dominated in the biomass sludge. But currently six recognized genera of *Candidatus* Scalindua, *Ca.* Kuenenia, *Ca.* Jettenia, *Ca.* Anammoxoglobus, *Ca.* Anammoximicrobium and *Ca.* Brocadia were discovered for anammox bacteria with significantly distinct kinetic and physiological characteristics including maximum specific growth rate, biomass yield coefficient and affinity constants (Ali et al., 2015; Zhang et al., 2017). Thirdly, most of studies (Ni et al., 2012, 2009) skipped a systematic model validation for independent scenarios with varying nitrogen availability under anoxic conditions to explain plausible interactions of heterotrophic denitrifiers and anammox bacteria. There are also works succeeded to validate the biofilm models for variant scenarios, but only applicable for aerobic conditions and interactions between autotrophs (Liu et al., 2016a; Peng et al., 2016) or interactions of autotrophs-heterotrophs without anammox bacteria (Carlos Domingo-Félez et al., 2016).

In this work, the main aim is to apply a simplified mathematical biofilm model to estimate kinetics and physiological parameters of microbial groups during anammox and denitrification processes using a novel stoichiometric matrix. A novel stoichiometric matrix includes: (i) new coefficients based on Eq. 7. (ii) differentiation between kinetic and physiology of major groups of anammox species which were recognized in the granular biomass using a molecular cytogenetic assay. The focus will be on estimation of maximum growth rate ($\mu_{max}$) and decay rates (b) of dominant anammox bacteria. After model build-up, several independent batch tests are conducted. For this mean, first, most sensitive parameters must be ranked, and a sufficiently identifiable set of parameters will be selected for calibration. After the estimation of selected kinetics and physiological parameters, the model will be validated for various independent scenarios.

## 3.2 Materials and methods

### 3.2.1 Reactor operation and biomass collection

Four independent batch assays were performed in duplicate to evaluate the anammox and denitrifying activity in anammox biofilm under anoxic condition. The origin of the biomass collected for the batch assays was granular biofilm from a full-scale landfill leachate treatment plant running for a central waste disposal site (AGR Group, Herten, Germany). In that plant, silica gel activated carbon granules of 1 to 3 mm with the density of 8 g $L^{-1}$ and the electrical conductivity (EC) of 14 mS were used as adsorbent material as well as a moving bed nucleus for the attachment and growth of biofilm. Synthetic wastewater was used as feed for the 12 h batch experiments. The full medium composition and the trace element solution used for all batch tests was previously described (Ke, 2014). The batch assays were performed in lab-scale for 300 mL reactor with a bulk liquid volume of 270 mL and the solid biomass volume amounting to 30 mL. The granular sludge concentration was ranging from 3.4 to 4.1 g $L^{-1}$ of mixed liquor volatile suspended solids (MLVSS) (Azari et al., 2017a).

The first assay was used for model calibration including sensitivity analysis, parameter identifiability analysis and parameter estimation. The three remaining batch assays were implemented to validate the model performance using estimated parameters. The concentration of nitrogen components and glucose as main organic carbon source in synthetic wastewater solution varies for each batch test and it is described in Table 4.

Table 4: Medium composition in medium solutions used for four independent batch assays. Each assay represents a modelling scenario.

| | Batch assay (scenario) | | | |
| | 1 | 2 | 3 | 4 |
| Component / propose | Calibration | Validation | Validation | Validation |
|---|---|---|---|---|
| NH$_4$Cl [mgN L$^{-1}$] | 61 | - | - | 59.0 |
| NaNO$_2$ [mgN L$^{-1}$] | 62 | - | - | - |
| NaNO$_3$ [mgN L$^{-1}$] | - | 62.5 | 58.5 | 60 |
| CaCl$_2$ [mg L$^{-1}$] | 135.9 | 135.9 | 135.9 | 135.9 |
| K$_2$HPO$_4$ [mg L$^{-1}$] | 34.8 | 34.8 | 34.8 | 34.8 |
| MgSO$_4$ [mg L$^{-1}$] | 146.5 | 146.5 | 146.5 | 146.5 |
| NaHCO$_3$ [mg L$^{-1}$] | 420.0 | 420.0 | 420.0 | 420.0 |
| Glucose [mgCOD L$^{-1}$] | - | - | 75 | - |
| Trace element [mg L$^{-1}$] | 1.0 | 1.0 | 1.0 | 1.0 |

Batch assay 1 was done in the presence of nitrite and ammonium and absence of nitrate and organic substrate. Afterwards, assays 2 and 3 were conducted in presence of nitrate to validate the performance of the model in predicting the activity of heterotrophs to remove nitrate in absence and presence of glucose as an organic substrate. Assay 4 was performed for model validation during simultaneous ammonium and nitrate removal in presence of ammonium and nitrate but in the absence of nitrite and organic substrate. The temperature during all assays was $32\pm1°C$, under very low dissolved oxygen (DO) concentration of approx. 0.04 mg $O_2$ $L^{-1}$ and pH value of 7.5. Hourly data of inorganic nitrogen components i.e. $NH_4$-N, $NO_2$-N and $NO_3$-N were measured photo-metrically for all batch assays starting from beginning time step according to the German standards using commercial cuvette tests (Dr. Lange, Hach, Germany).

### 3.2.2  Q-FISH for detection and quantification of dominant microbial groups

Quantitative fluorescence in situ hybridisation (qFISH) as an inexpensive semi-quantitative cytogenetic method was also used to investigate microbial populations present in the biomass. The existence of anammox *Planctomycetes* as major bacterial group in the same biomass was previously proven using FISH by fluorescently-labelled oligonucleotide probe Amx368 and AMX820 targeting all anammox *Planctomycetes*. It was also claimed that two major group of anammox bacteria found in this plant are from *Ca.* Brocadia and *Ca.* Scalindua genera. (Azari et al., 2017a,b; Ke et al., 2015). Thanks to the previous studies, collection of biofilm samples for experimental assays was performed to further check the relative abundance of anammox species particularly *Ca.* Brocadia anammoxidans and *Ca.* Scalindua sp. which were previously found as dominant species. The granular sludge was at steady state for experimental conditions. The list of general and specific probes for anammox bacteria, ammonia-oxidizing bacteria (AOB) and nitrite oxidizing bacteria is shown in Table 5. For the visualization of all cells (active cells and inactive or inert cells), DAPI (4',6-diamidino-2-phenylindole) was used. For the visualization of most of domain active bacteria, FISH using S-D-Bact-0338-a-A-18 so called EUB338 probe covering 90 % of domain bacteria was used (Biomers.net GmbH, Ulm, Germany). The FISH slides were visualized under light fluorescence microscopy (Axio imager 2, Carl Zeiss AG, Jena, Germany). For each sample, at least twenty random fields of view (FOV) were imaged, each sample in duplicate to determine the average relative abundance of each microbial group. The estimation of the area of average inactive (dead) was done by calculating the stained area taken by DAPI targeted cells subtracted by the area stained by the EUB 338 probe. Quantification of average active relative abundance of microbial species was done by calculating the stained area taken by fluorescence targeted cells from specific probes compared to the area stained by the EUB 338 probe targeting most of alive (active) bacteria in the biofilm. For inactive relative abundance, the average inactive area was compared with the area stained by EUB338 probe. The entire optimized procedure for sample preparation, qFISH, image acquisition and digital image analysis is described in the supplementary materials A. Calculated relative abundance values using qFISH, were used for initiating the biomass composition range.

Table 5: 16 S rRNA-targeted oligonucleotide probes used in this study for qFISH

| robes | Sequence (5 -3) | Specificity | Dye | References |
|---|---|---|---|---|
| UB338 | GCTGCCTCCCGTAGGAGT | Most bacteria | FITC | (Zarda et al., 1991) |
| ca1309 | TGGAGGCGAATTTCAGCCTCC | genus *Candidatus* Scalindua | Cy3 | (Schmid et al., 2000, 2007) |
| S820 | TAATTCCCTCTACTTAGTGCCC | *Ca.*Scalindua wagneri and *Ca.*Scalindua sorokinii | Cy5 | (Kartal et al., 2006) |
| mx820 | GCTGCCACCCGTAGGTGT | *Ca.* Brocadia anammoxidans, *Ca.* Kuenenia stuttgartiensis | Cy3 | (Kartal et al., 2006) |
| an162 | CGGTAGCCCCAATTGCTT | *Ca* Brocadia anammoxidans | Cy5 | (Arrojo et al., 2008) |
| mx368 | CCTTTCGGGCATTGCGAA | all anammox bacteria | Cy3 | (Kartal et al., 2006) |
| pr820 | AAACCCCTCTACCGAGTGCCC | *Ca.* Anammoxoglobus propionicus | Cy5 | (Kartal et al., 2007) |
| so190 | CGATCCCCTGCTTTTCTCC | Proteobacterial AOB | Cy3 | (Mobarry et al., 1997) |
| T3 | CCTGTGCTCCATGCTCCG | Nitrobacter spp | Cy3 | (Mobarry et al., 1997) |
| spa 662 | GGAATTCCGCTCTCCTCT | genus Nitrospira | Cy3 | (Daims et al., 2000) |

### 3.2.3 Mathematical modeling approach

The application and importance of Activated Sludge Model (ASM) for nitrogen removal was evaluated by conducting simulations under both steady-state and dynamic conditions (Hiatt and Grady, 2008). The Activated Sludge Model No. 1 (ASM1) by the task group of the International Water Association (IWA) was used (Henze, 2000; Henze et al., 1987). ASM1 model was further modified with the subsequent conversion of nitrate into nitrogen via nitrite (Ostace et al., 2011) and one-step anaerobic ammonia oxidation (anammox) process (Dapena-Mora et al., 2004). The biofilm model in software AQUASIM 2.1 (EAWAG,

Switzerland) was applied. A definition of the components used in the SAD model is given in Table 6.

Table 6: Definition of particulate and dissolved components used in the model

| Number | Component | Concentration | Unit |
|--------|-----------|---------------|------|
| Model dissolved components | | | |
| 1 | $S_{O2}$ | Dissolved oxygen | $gO_2\ m^{-3}$ |
| 2 | $S_S$ | Readily degradable organic substrate | $gCOD\ m^{-3}$ |
| 3 | $S_{NH4}$ | Ammonium nitrogen | $gN\ m^{-3}$ |
| 4 | $S_{NO2}$ | Nitrite nitrogen | $gN\ m^{-3}$ |
| 5 | $S_{NO3}$ | Nitrate nitrogen | $gN\ m^{-3}$ |
| 6 | $S_{N2}$ | Dissolved dinitrogen gas | $gN\ m^{-3}$ |
| Model particulate components | | | |
| 7 | $X_{AN,Sca}$ | Anammox bacteria: $Ca.$ Scalindua sp. | $gCOD\ m^{-3}$ |
| 8 | $X_{AN,Bro}$ | Anammox bacteria: $Ca.$ Brocadia anammoxidans | $gCOD\ m^{-3}$ |
| 9 | $X_H$ | Heterotrophic bacteria | $gCOD\ m^{-3}$ |
| 10 | $X_S$ | Slowly degradable organic substrate | $gCOD\ m^{-3}$ |
| 11 | $X_I$ | Inert, non-biodegradable organics | $gCOD\ m^{-3}$ |

Heterotrophic biomass and autotrophic anammox biomass are generated by growth on the readily biodegradable substrate ($S_S$) or by growth on ammonia nitrogen ($S_{NH4}$). The biomass is lost via the decay process where it is converted to (1) particulate non-biodegradable inert ($X_I$) and (2) particulate slowly biodegradable substrate ($X_S$). The latter could return to the process and be used by the remaining organisms as new source of readily biodegradable substrate ($S_s$) through hydrolysis. Monod kinetics has been used to express the dependencies of the growth rates to the substrates' concentration. The full lists of revised process kinetic rate equations including two group of major anammox species with distinct parameters for $Ca.$ Brocadia anammoxidans, $Ca.$ Scalindua sp. (based on Table 7) and heterotrophic denitrifiers and the revised stoichiometric matrix based on Eq. 7 are in Table C1 and C2 of Supplementary Materials respectively.

For the anammox growth rates equation, two definitions have been assessed. The first model definition considers the inhibition constant for oxygen for anammox organisms according to (A Dapena-Mora et al., 2004). But the second definition is the form without the term describing the oxygen inhibition for anammox growth. In this way, it can be determined how the oxygen inhibition term can influence modeled results under oxygen limited/anoxic conditions. Ammonia oxidizing bacteria (AOB) and nitrite oxidizing bacteria (NOB) were checked by qFISH but not included in the structure of the model due to anoxic conditions. This assumption complies with previous findings (Richardson et al., 2009), showing that when faced with a shortage of oxygen, bacterial species predominantly use nitrate and nitrite to support respiration via the process of two stage denitrification defined in the model and nitrite via the process of anammox.

Table 7: Distinct range and values for physiological and kinetic characteristics of various anammox species (Ali et al., 2015; Zhang et al., 2017)

| Species/ Parameter | *Ca.* Jettenia caeni | *Ca.* Brocadia sinica | *Ca.* Brocadia anammoxidans | *Ca.* Kuenenia stuttgartiensis | *Ca.*Scalindua sp. |
|---|---|---|---|---|---|
| Yield (Y) (gCOD gN$^{-1}$) | 0.128 | 0.144 | 0.16 | 0.16 | 0.068 |
| $\mu_{max}$ (h$^{-1}$) | 0.002-0.006 | 0.004-0.014 | 0.0025-0.0027 | 0.0026-0.0096 | 0.002-0.007 |
| Affinity constant (K) | | | | | |
| NH$_4^+$ (gN m$^{-3}$) | 0.24±0.06 | 0.39±0.056 | 0.07 | 0.07 | 0.042 |
| NO$_2^-$ (gN m$^{-3}$) | 0.49±0.013 | 0.48±0.3 | 0.07 | 0.05 | 0.0063 |

### 3.2.4 Definition of the biofilm model and geometrical structure

Granular sludge used in this study consists of dense and spherical biofilms with the diameter ranging from 0.1 to 20 mm (Ke et al., 2015). Hence, to avoid complexity, the model encompasses a simplified scheme to characterize homogenous spherical biofilms (Fig. 21). From the biophysical point of view, a multispecies one-dimensional simple biofilm model in a confined batch reactor with a constant volume of 300 mL was considered. The biofilm model is assumed to have three layers: solid biofilm matrix, boundary layer (also called pore water) and bulk liquid part (Fig. 21). The detachment velocity was not considered due to the short simulation period and the rate porosity within the biofilm surface was set on a constant level of 0.4. The mass balance and continuity equations were described due to (Reichert, 1998, 1994) but no diffusive mass transport of biomass within the solid biofilm matrix was considered.

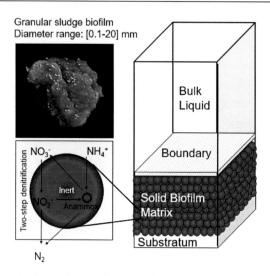

Figure 21: Proposed scheme for one-dimensional multispecies granular biofilm model for anoxic simulation of anammox & denitrification

Boundary layer resistance (also called as mass transfer coefficient ($k_L$)) is equal to:

$$k_L = \frac{D_i}{L_L}$$

(Eq. 15)

Where ($D_i$) is the diffusion coefficient for a particulate or dissolved compound (i) and $L_L$ is the thickness of the external boundary layer thickness (Horn and Lackner, 2014). The biofilm and bulk liquid compartments are in contact and exchange solutes only by diffusion. To determine the loss of biofilm surface area between multiple pieces of granules (considered as spheres) the effective biofilm surface area was calculated. The Eq. 16. gives the effective area of the biofilm which is the remaining surface area of the biofilm interacting with substrate:

$$A_{eff} = c. A_{tot}$$

(Eq. 16)

Where $A_{tot}$ is the total surface area of all granular biofilm excluding the lost surface and c which represents to the empirical correction factor (0 to 1) for the surface area based on contact points of granular spheres. $A_{tot}$ itself is calculated from Eq. 17:

$$A_{tot} = 4\pi. \, n \, (LF + r_r)^2$$

(Eq. 17)

$A_{tot}$ is in relation to the number of contact granules (n), the radius ($r_r$) of the of single spherical particle (one support cell) varying between $10^{-8}$ m to $10^{-5}$ m and the actual thickness of the biofilm (LF) calculated by the model. The number of pieces of granules (n) assuming the homogenous size (diameter) for all granular spheres is as Eq. 18:

$$n \cong \frac{3\,V_{Biomass}}{4\pi r_{sp}^3} \qquad\qquad\qquad\text{(Eq. 18)}$$

Where $r_{sp}$ is the initial biofilm thickness at the start of experiments. Throughout simulations the total biomass density in the biofilm solid matrix which were kept constant during the simulation periods (rho). Selected values of biofilm, mass transfer, diffusion and geometrical parameters were obtained from a previous model for the same biomass (Azari et al., 2017b) (Table C3 of Supplementary Materials).

### 3.2.5 Statistical analysis

The local sensitivity analysis and the analysis of error contribution of parameters were performed on model results for the calibration scenario. The absolute-relative sensitivity function in AQUASIM was chosen to identify the most important parameters influencing on prediction of three output variables including $NH_4$-N, $NO_2$-N and $NO_3$-N (Reichert, 1998). This criterion measures the absolute change in y (one output variable) for a 30 % change in p as an assessed parameter as displayed in Eq. 19. To avoid a significant impact of non-linear model behavior only 30 % of change was adjusted. The sum of sensitivity functions (SF) is used to compare the effect of change in each parameter with regards to the sum of output variables.

$$\delta_{y,p}^{AR} = p\,\frac{\partial y}{\partial p} \qquad\qquad\qquad\text{(Eq. 19)}$$

In which $\delta_{y,p}^{AR}$ is the function of absolute-relative sensitivity for each parameter, Y is selected output function ($S_{NH4}$, $S_{NO2}$ or $S_{NO3}$) and p indicates a single parameter considered in the sensitivity analysis.

The error contribution of each parameter on total uncertainty of the model prediction has been calculated as an uncertainty of model parameters which can propagate to the uncertainty of model results. The average error contribution of each parameter was calculated for all model output variables. The simple error propagation method was defined as the linearized propagation of standard deviations of uncorrelated parameters. The error propagation formula used a linearized model in which $p_i$ is the uncertain model parameter, $\sigma_{pi}$ is their standard deviation of these parameter, y $(p_1,...,p_m)$ is the solution of the model equations for a given variable at a given location and time for various sets of parameter pi, and $\sigma_y$ is the approximated standard deviation of output variable defined by Eq. 20. The character "m" is the number of iterations in which parameter p has been changed within the defined range and it was set to perform 100 iterations.

$$\sigma_y = \sqrt{\sum_{i=1}^{m} \sigma_{p_i}^2 \left(\frac{\partial y}{\partial p_i}\right)^2}$$ (Eq. 20)

From sufficiently sensitive parameters, an arbitrary parameter subset k was selected and the identifiability analysis of parameters was done by measuring the collinearity index $\lambda_K$. Collinearity index is a tool for the analysis of parameters interdependencies and quantification of the degree of approximate linear dependence of the sensitivity functions of the parameters of the selected subset (Brun et al., 2001). The procedure for calculation of collinearity index is explained elsewhere (Brun et al., 2002). If $\lambda_K$ <10, the combination of selected parameters is identifiable enough while $\lambda_K$ >10 means that this parameter subset is poorly identifiable. A large $\lambda_K$ means that changes in model results induced by the change in one parameter (of the subset) can be approximately compensated by appropriate changes in the other parameters of the subset (Mieleitner and Reichert, 2006).

Upon choosing the identifiable parameters in the model, the parameter estimation was implemented for the calibration scenario after several iterations of parameters tuning including manual and automated calibration by maximizing R-squared ($R^2$) and minimizing the chi-square ($\chi^2$) values between the model data and the actual data using the Simplex nonlinear parameter estimation algorithm (Ficken, 2015) in AQUASIM v2.1 (Reichert, 1998, 1994). Selected kinetic and stoichiometric parameters giving the best agreement between hourly simulation and observation data points for the calibration scenario were selected for the rest of the model validation simulations. R-squared was also used to evaluate the model performance for validation scenarios. For uncertainty analysis, a simple linear regression equation was assumed to model the nitrogen loss using the observation data points. The confidence regions of parameters including intercept and slope were assumed to be linear and a 95 % confidence interval region around the slope of a regression line was constructed by calculating the inverse of the cumulative normal distribution function for an assumed value of x, and a supplied distribution mean and corresponding standard deviation of y values. In this case, x-values will be the time steps in hour and y-values will be the relevant observed nitrogen components' concentration.

## 3.3 Results and discussion

### 3.3.1 Initialization of the biomass composition

From qFISH results, the relative abundance out of total active cells was of 35.5 % to 61.9 % for *Ca.* Brocadia anammoxidans (Amx820 and Ban162 probes), of 3.4 to 11.8 % for the genus *Ca.* Scalindua sp. (Sca1309), of 1.3 to 5.4 % for AOB (Nso190 probe) and of less than 4.5 % for NOB (Ntspa662 and NIT3 probes). Other anammox species were negligible for considering in the model. The inert matter fraction out of total cells was estimated as 15.6 to 28.6 %. Due to the very low oxygen concentration in the reactor the AOB and NOB fractions were skipped. The mean values were selected as initial fractions and the remaining fraction was attributed to heterotrophs. Selected FISH images of *Ca.* Brocadia and *Ca.* Scalindua as two major genera in the same biomass (but from different sampling date) are presented in last chapter.

### 3.3.2 Sensitivity and error contribution analysis

Most sensitive parameters (15 out of 30 considered parameters) having the largest influence on the results (Fig. 22a) The results illustrate that the model prediction is highly sensitive to the biofilm density (rho) as well as yield coefficients (Y) and maximum specific growth rates ($\mu$) of denitrifying and anammox bacteria. Fig. 2B elucidates that 9 parameters contributed more than 91 % of the model error while the remaining 21 parameters amounted to less than 9 % of the total error. Among top parameters, the radius ($r_r$) of the of single spherical particle (support cell) is ranked first and $L_L$ as external mass transfer boundary layer thickness is the second contributor to the total error. Since in this biofilm model, physical processes like mass transfer reactions and diffusion equations are based on Fick's Law (Horn and Lackner, 2014), substrates in the bulk liquid phase can be transported to the biofilm, and a mass transfer boundary layer ($L_L$) is assumed around granules. Definition of $L_L$ clarifies that near the granule the diffusion becomes critical for moving the solute toward or away from the biofilm surface. In the biofilm system, as $L_L$ increases, the fluid flow is reduced, and the diffusion distance is increased. Therefore, the highest contribution of $L_L$ and $r_r$ imply the importance of mass transfer phenomena in the biofilm. However, diffusivity coefficients of each substrates and $r_{sp}$ (initial average radius size of one granule) do not have that significant error contribution compared to $r_r$ and $L_L$.

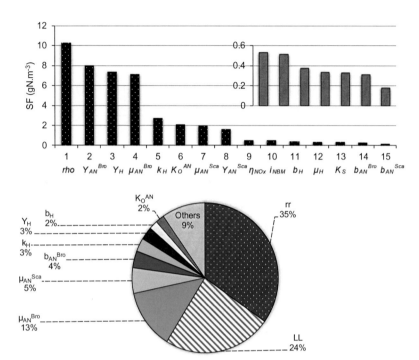

Figure 22: Sum of sensitivity analysis functions (SF) per 15 ranked parameters (top) and the ranked error contribution out of total model uncertainty (buttom).

### 3.3.3 Parameter identifiability analysis

To assess the overall identifiability of the most sensitive parameters, collinearity indices were calculated and presented in Table 8. The results show whether changes in three output variables i.e. $NH_4$-N, $NO_2$-N and $NO_3$-N concentrations, caused by changes in one parameter can be compensated by changes in other parameters. The analysis was done twice before the calibration with initial arbitrary values for parameters ($\theta_{ini}$) and after the calibration ($\theta_{end}$). Among 15 sensitive parameters, for biofilm density (rho) in terms of chemical oxygen demand (COD) a value equal to 148,000 g COD $m^{-3}$ was set based on the previous studies (Azari et al., 2017b) and yield coefficients of *Ca. Brocadia anammoxidans* and *Ca. Scalindua* sp., were obtained from Table 7. For the maximum specific growth rate parameters of *Ca.* Brocadia anammoxidans *and Ca.* Scalindua sp., the appropriate range was used from Table 7 for further parameter tuning. For remaining sensitive parameters, a typical range of values was considered from literatures (Table C4 of supplementary materials). The final set of 12 remaining kinetic and physiological parameters suggested for parameter estimation, resulted in a collinearity index ($\lambda$) of 5.6 for $\theta_{ini}$ and 7.3 for $\theta_{end}$ (Table 7). This means that this set of parameters are sufficiently identifiable since $\lambda < 10$.

The final list of stoichiometric and kinetic parameters leading to obtain the best efficiency for the calibration scenario with lowest Chi-squared and highest $R^2$ and the sources are presented in Table 8.

Table 7: Collinearity index ($\lambda$) for selected subsets of parameters

| Set number | Set size | $\lambda(\theta_{ini})$ [1] Before the calibration | $\lambda(\theta_{end})$ [2] After the calibration | List of parameters in the set |
|---|---|---|---|---|
| 1 | 2 | 1.9 | 2.4 | $rho, Y_{AN}^{Bro}$, |
| 2 | 3 | 2.4 | 6.7 | $rho, Y_{AN}^{Bro}, Y_H$ |
| 3 | 4 | 9.3 | 9.1 | $rho, Y_{AN}^{Bro}, Y_H, \mu_{AN}^{Bro}$ |
| 4 | 5 | 52.0 | 39 | $rho, Y_{AN}^{Bro}, Y_H, \mu_{AN}^{Bro}, k_H$ |
| 5 | 6 | 40.1 | 34.2 | $rho, Y_{AN}^{Bro}, Y_H, \mu_{AN}^{Bro}, k_H, k_O^{AN}$ |
| 6 | 7 | 13.4 | 24.2 | $rho, Y_{AN}^{Bro}, Y_H, \mu_{AN}^{Bro}, k_H, k_O^{AN}, \mu_{AN}^{Sca}$ |
| 7 | 8 | 12.9 | 13.3 | $rho, Y_{AN}^{Bro}, Y_H, \mu_{AN}^{Bro}, k_H, k_O^{AN}, \mu_{AN}^{Sca}, Y_{AN}^{Sca}$ |
| 8 | 9 | 8.2 | 10.6 | $rho, Y_{AN}^{Bro}, Y_H, \mu_{AN}^{Bro}, k_H, k_O^{AN}, \mu_{AN}^{Sca}, Y_{AN}^{Sca}, \eta_{NOX}$ |
| 9 | 10 | 10.0 | 9.1 | $rho, Y_{AN}^{Bro}, Y_H, \mu_{AN}^{Bro}, k_H, k_O^{AN}, \mu_{AN}^{Sca}, Y_{AN}^{Sca}, \eta_{NOx}, i_{NBM}$ |
| 10 | 11 | 8.77 | 5.8 | $rho, Y_{AN}^{Bro}, Y_H, \mu_{AN}^{Bro}, k_H, k_O^{AN}, \mu_{AN}^{Sca}, Y_{AN}^{Sca}, \eta_{NOx}, i_{NBM}, b_H$ |
| 11 | 12 | 8.1 | 8.7 | $rho, Y_{AN}^{Bro}, Y_H, \mu_{AN}^{Bro}, k_H, k_O^{AN}, \mu_{AN}^{Sca}, Y_{AN}^{Sca}, \eta_{NOx}, i_{NBM}, b_H, \mu_H$ |
| 12 | 13 | 7.1 | 7.7 | $rho, Y_{AN}^{Bro}, Y_H, \mu_{AN}^{Bro}, k_H, k_O^{AN}, \mu_{AN}^{Sca}, Y_{AN}^{Sca}, \eta_{NOx}, i_{NBM}, b_H, \mu_H, K_S$ |
| 13 | 14 | 6.8 | 7.5 | $rho, Y_{AN}^{Bro}, Y_H, \mu_{AN}^{Bro}, k_H, k_O^{AN}, \mu_{AN}^{Sca}, Y_{AN}^{Sca}, \eta_{NOx}, i_{NBM}, b_H, \mu_H, K_S, b_{AN}^{Bro}$ |
| 14 | 15 | 6.9 | 7.2 | $rho, Y_{AN}^{Bro}, Y_H, \mu_{AN}^{Bro}, k_H, k_O^{AN}, \mu_{AN}^{Sca}, Y_{AN}^{Sca}, \eta_{NOx}, i_{NBM}, b_H, \mu_H, K_S, b_{AN}^{Bro}, b_{AN}^{Sca}$ |
| 15 [3] | 12 | 5.6 | 7.3 | $Y_H, \mu_{AN}^{Bro}, k_H, k_O^{AN}, \mu_{AN}^{Sca}, \eta_{NOx}, i_{NBM}, b_H, \mu_H, K_S, b_{AN}^{Bro}, b_{AN}^{Sca}$ |

[1] $\theta_{ini}$ refers to the arbitrary initial parameter values.

[2] $\theta_{end}$ refers to the optimized final set of parameters after carrying out the calibration.

[3] The final set of twelve potentially identifiable ($\lambda < 10$) among top 15 sensitive parameters, which selected for the calibration and tuning procedure. This set gives the lowest collinearity index $\lambda(\theta_{ini}) = 5.6$. Note that $Y_{AN}^{Bro}$ and $Y_{AN}^{Sca}$ were selected from Table 7 and rho was selected from previous studies. Therefore, only 12 out of 15 sensitive parameters must be calibrated.

Table 8: Final list of selected values for kinetic, physiologic and stoichiometric parameters

| Parameter | Definition | range | Unit |
|---|---|---|---|
| $\mu_{AN}^{Bro}$ [1] | maximum growth rate of $Ca.$ Brocadia anammoxidans | 0.0025 | $h^{-1}$ |
| $\mu_{AN}^{Sca}$ [1] | maximum growth rate of $Ca.$ Scalindua | 0.0048 | $h^{-1}$ |
| $b_{AN}^{Bro}$ [1] | decay rate coefficient of $Ca.$ Brocadia anammoxidans | 0.0003 | $h^{-1}$ |
| $b_{AN}^{Sca}$ [1] | decay rate coefficient of $Ca.$ Scalindua | 0.00026 | $h^{-1}$ |
| $\gamma_{AN}^{Bro}$ [2] | anoxic yield coefficient for $Ca.$ Brocadia anammoxidans | 0.16 | gCOD gN$^{-1}$ |
| $\gamma_{AN}^{Sca}$ [2] | anoxic yield coefficient for $Ca.$ Scalindua | 0.068 | gCOD gN$^{-1}$ |
| $K_{NH4}^{AN,Bro}$ [2] | $S_{NH4}$ affinity constant for $Ca.$ Brocadia anammoxidans | 0.07 | gN m$^{-3}$ |
| $K_{NH4}^{AN,Sca}$ [2] | $S_{NH4}$ affinity constant for $Ca.$ Scalindua | 0.042 | gN m$^{-3}$ |
| $K_{NO2}^{Bro}$ [2] | $S_{NO2}$ affinity constant for $Ca.$ Brocadia anammoxidans | 0.07 | gN m$^{-3}$ |
| $K_{NO2}^{Sca}$ [2] | $S_{NO2}$ affinity constant for $Ca.$ Scalindua | 0.0063 | gN m$^{-3}$ |
| $K_O^{AN}$ [1], [3] | $S_{O2}$ inhibiting coefficients for all anammox bacteria | 0.1 | gO$_2$ m$^{-3}$ |
| $Y_H$ [1] | anoxic yield coefficient for heterotrophic denitrifiers | 0.45 | gCOD g$^{-1}$COD |
| $\mu_H$ [1] | maximum growth rate of heterotrophic denitrifiers | 0.25 | $h^{-1}$ |
| $b_H$ [1] | decay rate coefficient of heterotrophic denitrifiers | 0.014 | $h^{-1}$ |
| $k_H$ [1] | hydrolysis rate constant | 0.05 | $h^{-1}$ |
| $K_x$ [4] | hydrolysis saturation constant | 0.03 | gCOD |
| $K_s$ [1] | $S_S$ affinity constant for heterotrophic denitrifiers | 20 | gCOD m$^{-3}$ |
| $K_{NH4}^H$ [4] | $S_{NH4}$ affinity constant for heterotrophic denitrifiers | 0.01 | gN m$^{-3}$ |
| $K_{NOx}^H$ [4] | $S_{NOx}$ affinity constant for heterotrophic denitrifiers | 0.5 | gN m$^{-3}$ |
| $K_O^H$ [4] | $S_{O2}$ inhibiting coefficients for heterotrophic denitrifiers | 0.2 | gO$_2$ m$^{-3}$ |
| $\eta_{NOx}$ [4] | heterotroph anoxic reduction factor | 0.8 | — |
| $i_{NBM}$ [1] | nitrogen content of biomass | 0.078 | gN g$^{-1}$COD |
| $i_{NXI}$ [4] | nitrogen content of inert biomass $X_I$ | 0.06 | gN g$^{-1}$COD |
| $f_i$ [4] | fraction of biomass decaying into inert biomass | 0.08 | gCOD g$^{-1}$COD |

[1] Source of values are from this study after the calibration of 12 identifiable parameters (see Table 7) based on valid ranges defined in Table 3 and Table C3,4.

[2] Source of values are selected from Table 3 for $Ca.$ Brocadia and $Ca.$ Scalindua.

[3] It is assumed that: $K_{O2}^{AN} = K_{O2}^{AN,Bro} = K_{O2}^{AN,Sca}$.

[4] Source of values are given elsewhere (Azari et al., 2017b).

### 3.3.4 Model calibration

Correlation coefficients of $R^2$ was calculated between the modeled and measured data points for the calibration scenario for two models with and without oxygen inhibition term ($K_O^{AN}$). Both models address a good agreement between measured and predicted values. However, the model with oxygen inhibition term showed a better goodness-of-fitness index ($R^2$) equal to 0.95, 0.96 and 0.67 for NH$_4$-N, NO$_2$-N and NO$_3$-N, respectively, while the $R^2$ values for the model skipping the oxygen inhibition factor are 0.90, 0.95 and 0.6 for NH$_4$-N, NO$_2$-N and NO$_3$-N, respectively (Fig. 23 and Table 9). This comparison helps to understand that due to the oxygen limited conditions, the oxygen inhibition term was not significantly influencing on nitrite and ammonium prediction, but the impact on nitrate prediction was notable. The effect of DO and a DO inhibition term on the simulation results would be more pronounced if there would be more variation in the DO concentrations. However, in statistical point of view the simplified model tends to overestimate the nitrogen components because there is no inhibition term considered for the growth of anammox due to oxygen concentration. Therefore, the model with oxygen inhibition term is still favorable and will be considered for validation.

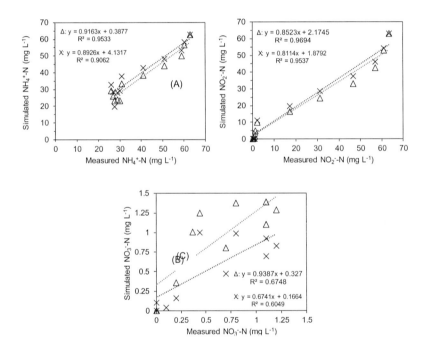

Figure 23. Analysis of regressions for model calibration results for three output variables (A-C). Blue dotted lines and X symbols indicate the values of the simplified version of the model without the oxygen inhibition term and dotted red line and Δ symbols are the values of the main model considered with the oxygen inhibition term in anammox growth rate.

Table 9: Calculated statistics for two model definitions for three output variables for calibration

| Statistical Index | | R-squared index ($R^2$) [-] | | Chi-square ($\chi2$) [mgN $L^{-1}$]$^2$ | |
|---|---|---|---|---|---|
| Model definition/ nitrogen component | Inhibited by oxygen | No oxygen inhibition term for anammox | Inhibited by oxygen | No oxygen inhibition term for anammox |
| $NH_4$-N | 0.95 | 0.9 | 447 | 501 |
| $NO_2$-N | 0.96 | 0.95 | 414 | 466 |
| $NO_3$-N | 0.67 | 0.6 | 518 | 532 |

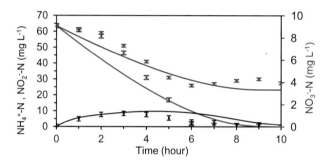

Figure 24: Time dependent calibrated results for three output variables for assay 1 during half-day simulation. Blue lines and dots indicate for $NH_4$-N simulation and observations respectively. Red line and dots indicate for $NO_2$-N simulation and observations. Black line and dots indicate for $NO_3$-N simulation and observations. Error bars are shown based on standard deviations of multiple chemical measurements.

The comparison between measured $NH_4$-N and the predicted concentration of $NH_4$-N by the model indicated that a significant loss of $NH_4$-N during the first batch assay was simulated correctly. In Fig.4, a significant decline in $NH_4$-N concentration within the first 7 hours due to the activity of anammox bacteria was simulated. During the time step 6 to 7 hour, the nitrite is entirely consumed because the supplied $NO_2$-N/$NH_4$-N ratio in the first batch assay is equal to 1.016 (Table 3) which is lower than the anammox process theoretical stoichiometric ratio of 1.146 (Eq. 7). Therefore, no further ammonium biodegradation is predicted after 7 hours and the slight release of ammonium due to the decay of anammox bacteria can be observed from the time steps 7 to 10 h. After the time step 7 h, the prediction follows the results of the calibration (Fig. 24), outlining the release of ammonium concentration due to the decay of anammox autotrophs.

Elevated nitrate occurred from time step 0 to 4 h and after 6 h, the nitrate level declined. Until 4 h after the start of the assay 1, the anammox bacteria outcompete the denitrifiers so that nitrate concentration raises and at the time step 5 h, the $NO_3$-N concentration is spiked to > 6 mgN $L^{-1}$. The model can reliably simulate both anammox growth and decay process and after 6h because there is no electron acceptor for growth of anammox in the system, nitrite declined in the system. Therefore, the remaining produced nitrate will be consumed by heterotrophic denitrifies. This decline continuous due to the activity of denitrifiers outcompeting the anammox bacteria. Besides, the full process of anammox in which one mole of $NH_4^+$ and 1.146 mole of $NO_2^-$ produce 0.161 mole of $NO_3^-$ is showed elsewhere (Yao et al., 2015). This fact is clearly showed by the model for nitrate production prediction during the activity of anammox bacteria.

### 3.3.5  Model validation

Experimental data of three independent batch assays (scenarios 2, 3 and 4) defined in Table 1 were compared versus simulation. The model can simulate the variation trends for all scenarios and predicted values fall within the 95 % interval regions. In the scenarios 2 (Fig. 25a) and the scenario 3 (Fig. 25b), respectively single heterotrophic denitrifying activity in absence (Fig. 25a) and presence (Fig. 25b) of glucose as a readily biodegradable substrate ($S_s$) in the bulk liquid was validated. In Fig. 25a, nitrate reduction via endogenous denitrification is perfectly modeled within the 95 % confidence bounds. But Fig. 25b clarifies an underestimation in nitrate concentrations when glucose was added. Due to Fig 25b, by availability of glucose ($S_s$) as readily biodegradable substrate for heterotrophic organisms, the nitrate depletion rate is higher compared to Fig. 25a. The $NO_3^-$-N reduction rate within the first two hours during availability of glucose is 10 mgN $L^{-1}$ $h^{-1}$ while this value for endogenous activity is 5 mgN $L^{-1}$ $h^{-1}$. During the first 6 hours nitrate reduction rate for both the modeled and measured data is 5 mgN $L^{-1}$ $h^{-1}$ in absence of glucose ($S_s$). The nitrate reduction rate in presence of glucose for model and measurements is estimated around 7 mgN $L^{-1}$ $h^{-1}$. In general, nitrate reduction prediction results for both scenarios produced a strong $R^2$ between 0.93-9.96 and the simulation lines lays inside the calculated 95 % confidence regions. It must be noted that due to the limited number of data points, the confidence regions and correlation coefficient may not be reliable although the approach is statistically correct.

For the last scenario, results of the model validation are given in Fig. 25c, d. This scenario simulates the coexistence and simultaneous activity of anammox and denitrifying bacteria in the biofilm matrix. The model scarcely tends to overestimate $NH_4^+$-N variations over the last simulation time steps while the nitrate reduction prediction is strongly fitted to observed data. But the simulation line completely overlaid within the confidence interval range and in bot. h cases the correlation coefficient of 0.98 was calculated which confirms the good agreement between observation and simulation and with the defined criteria for a proper model performance (Fig. C2 of Supplementary Materials).

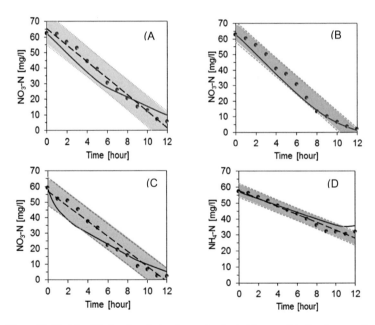

Figure 25: Model validation for nitrate nitrogen for the batch assay 2 in the absence of glucose and presence of nitrate (A), assay 3 in the presence of glucose and nitrate (B) and assay 4 in the presence of nitrate and ammonium (C, D). Dotted lines indicate the lower and upper limit of 95 % confidence regions and the black line is the main simulation line. The pink shaded area represents the 95 % confidence interval. Black spheres are measured values. Small spheres are observation points and black line is the predicted concentration by the model.

### 3.3.6 Kinetic and physiological characterization of Ca. Brocadia anammoxidans and Ca. Scalindua sp.

After identifying the dominant anammox species by qFISH, due to the range explained in Table 4 and Table C4 of Supplementary Materials, maximum growth rate and decay rate of *Ca.* Brocadia anammoxidans *and Ca.* Scalindua were estimated.

It was found that the maximum growth rate ( $_{max}$) at the temperature 32 ±1 °C for *Ca.* Brocadia anammoxidans is 0.0025 h$^{-1}$ (in compliance with classical value used for this species in aggregated form (Strous et al., 1998, 1999) while this value for *Ca.* Scalindua sp. is approximately two folds higher and equal to 0.0048 h$^{-1}$. The latter value for *Ca.* Scalindua sp. complies with recent findings suggesting a higher maximum growth rate of *Ca.* Scalindua genus for planktonic and aggregated cells which can be up to 0.007 h$^{-1}$ ((Zhang et al., 2017). In contrary, decay rates ($b_{AN}$) for *Ca.* Scalindua sp. was calibrated equal to 0.00026 h$^{-1}$ and for *Ca.* Brocadia anammoxidans equal to 0.0003 h$^{-1}$ (Dapena-Mora et al., 2004).

### 3.3.7 Theory of grey-box nitrogen elimination modeling

A grey-box modeling approach previously showed to reduce the uncertainty of nonlinear systems by using a combination of mechanistic (usually built from physical laws, conservation relations, and established) and black-box (empirical) model components (Romijn et al., 2008). The simplified model in this work illustrated how two types of empirical and mechanistic knowledge can be combined with existing experimental data. Several discrepancies observed between modeled and observed values could be accounted due to the complexity of kinetic behaviors of granular sludge under anoxic conditions which is not considered in this grey-box scheme (Fig. 26). For instance, underrating the $NO_3$-N concentration (Fig. 25a,b,c) might be due to overestimation of growth rate of denitrifies and due to underestimating the specific anammox activity. The latter is because the model under severe nitrite limiting conditions estimated the anammox activity rate to zero while in actual conditions, nitrite concentration is not physically zero. Because, nitrite can be also produced during the partial denitrification by nitrate reductase (NaR) which is not considered in this grey-box model. Therefore, during batch assays 2,3 and 4 in the absence of nitrite electron acceptor, $NO_2$-N concentration elevated from less than 0.1 mgN L$^{-1}$ to 2.2 mgN L$^{-1}$ after two hours and then dropped to zero (measured data and simulated data of nitrite is not shown). But in scenario 1 for the calibration, the introduction of nitrite into the system led to immediate consumption by anammox activity. Although the model considers two subsequent pathways for denitrification over nitrite (Fig. 26a) and denitrification over nitrate (Fig. 26b), it skips this possibility of partial denitrification of $NO_3^-$ to $NO_2^-$. Additionally, the production and survival of $NO_2^-$, NO and $N_2O$ from partial reductase are ignored, meaning that denitrification would proceed until $N_2$ and no pathways for prediction of NO or $N_2O$ release is proposed. Thus, due to underestimating the release of survived $NO_2$ during partial denitrification, anammox activity is underestimated which will lead to underestimating the nitrate values during the time steps two to five (Fig. 25a, b,c). Despite the model can satisfactorily predict general potential interactions between denitrifies and anammox bacteria but it is unable to model other metabolic pathways in detail. Calculated statistical indices, the subsequent conversion of nitrate into nitrogen via nitrite processes leading to production of molecular dinitrogen gas under nitrite (process B) or nitrate (process C) exposure are expressed satisfactorily. However, modeled processes in this model do not incorporate the enzymes nitrate reductase (NAR), nitrite reductase (NIR), nitric oxide (NO) reductase (NOR), and nitrous oxide reductase (N2OR) in details. Besides, nitrate reduction via dissimilatory nitrate reduction to ammonium (DNRA) and other aspects of the inorganic nitrogen metabolism could be also considered inside the grey box of the model.

The same fact is valid for anammox process (process C) since the catabolism and anabolism of *Planctomycetes* cannot be covered by any available model. The kinetics and stoichiometric evolutions of components inside grey boxes were not considered in the existing model structure (Fig. 26) leading to the model structure's uncertainty. Another uncertain aspect of this model was described by (Kartal et al., 2007), who investigated the possibility of dissimilatory nitrate reduction to ammonium by anammox bacteria when being physically purified in the anammox bacterium *Ca.* Kuenenia stuttgartiensis. This type of plausible dissimilatory pathway was not initiated in our kinetic model structure since qFISH results demonstrated that this organism showed a negligible relative abundance.

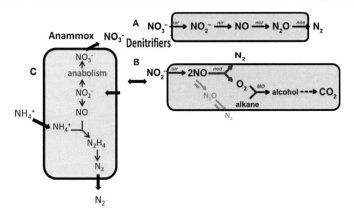

Figure 26: Proposed grey-box model for modeling nitrogen elimination during SAD process. Metabolic processes within grey boxes are not formulated in the model. Functional enzymes involved in this metabolism for denitrification of nitrate (A) are: nitrate ($NO_3^-$), nitrite ($NO_2^-$), nitric oxide (NO) and nitrous oxide ($N_2O$) reductases to dinitrogen gas ($N_2$) encoded by nar, nir, nor and nos gene clusters, respectively. Enzymes for denitrification of nitrite (B) are: nitrite, nitric oxide and nitrous oxide reductases encoded by nir, nor and nos gene clusters, respectively. Other pathways can be nitrite respiration (nir) and nitric oxide deoxygenation (nod). Consequently, monooxygenases (MO) are enzymes that catalyse the released oxygen atom from $O_2$ into an organic substrate such as alcohols. In anammox process (C), hydrazine ($N_2H_4$) is oxidized to $N_2$.


## 3.1   Conclusion

A mathematical model to simulate wastewater quality and biofilm growth was studied. The model incorporated two microbial groups for anammox bacteria and one group for heterotrophic denitrifying bacteria. Two major groups of anammox species included *Ca.* Scalindua sp. and *Ca.* Brocadia anammoxidans which were predominantly identified by qFISH. Different values and ranges of kinetic and physiological parameters for target groups were applied. Furthermore, an updated stoichiometry of anammox process based on a revisited equation given in 2014 was used. The main propose of the model was to study coexistence of heterotrophic denitrifiers and anammox bacteria under scenarios of varying nitrogen availability. After a comprehensive sensitivity analysis, an identifiable set of sensitive parameters was selected based on collinearity index and the parameters were tuned for the calibration scenario. The model was validated, and it could decently explain degradation of inorganic nitrogen species during anoxic conditions. The model performance briefed a good agreement between measured and predicted values with R-squared of more than 0.9 for all inorganic nitrogen components except for simulation of $NO_3$-N during the calibration scenario in which R-squared was equal to 0.67. Due to result, the decay rates and maximum growth rates of species belonging to *Ca.* Borcadia anammoxidans and *Ca.* Scalindua genus were estimated. Using modelling tools, it is proposed in this study that *Ca. Scalindua* sp. can have a higher maximum specific growth rate compared to *Ca.* Brocadia anammoxidans. The novel strategy given in this research is recommended to be followed in next model-based evaluation of anammox-based plants since due to previous researches, at least *Ca.* Jettenia caeni, *Ca.* Brocadia sinica, *Ca.* Brocadia anammoxidans, *Ca.* Kuenenia stuttgartiensis and *Ca.* Scalindua sp. showed distinct kinetic and physiological characteristics.

## 4 Model-based analysis of microbial consortia and microbial products in an anammox biofilm reactor

### 4.1 Introduction

Mathematical modeling of microbial community dynamics and interactions in mixed consortia is of great significance not only in ecological simulation but also for activated sludge models in wastewater treatment (Klimenko et al., 2016; Arashiro et al., 2017). In biological nitrogen removal processes, simulation of microbial population can help to develop a support tools to investigate the interactions between microorganisms. This step is also beneficial to improve calculation of substrate removal in bioreactors.

The anammox-based technologies are potential energy-positive alternatives to conventional nitrification-denitrification process especially for ammonia-rich wastewater treatment (Ma et al., 2016, Cao et al., 2016). Due to the slow growth of anammox bacteria, improvement of solids retention time (SRT) is required and therefore the application of anammox in biofilm reactors has been suggested as an approach (Lackner et al., 2014). Mathematical modeling of this type of anammox biofilm reactor could be an advantageous tool for the simulation, prediction and understanding of microbial community structure and biofilm characteristics. However, this is controversial due to present research gaps. First, current implications from mathematical modeling of biofilm reactors mainly focus on simulation of the substrates removal but little is known about modeling the active and inactive microbial groups regarding to dominance, abundance and biocoenosis. Also, in terms of dynamic modeling of extracellular polymeric substances (EPS) and soluble microbial products(SMP) with a focus on nitrogen-converting bacteria only few studies are available (Ni et al., 2012; Liu et al., 2016) while other studies have concluded that production of EPS components is linked to the formation of biofilms and granular sludge and the stability of bioreactor (Tan et al., 2017). In addition, model verification requires a systematic algorithm which has not been adequately addressed in current studies (Zhu et al., 2016). For example, based on the sensitivity analysis of previous works (Azari et al., 2017b,c), the model output results of such complex biofilm models are prone to parameters collinearity, numerical instability and minor changes in kinetic and stoichiometric parameters and sometimes initial parameters. Hence these values must be tuned and identified properly. Finally, among studies explaining models of anammox plants (Dapena-Mora et al., 2004; Ni et al., 2009,2014; Liu et al., 2017), few checked the model against good long-term experimental data, not only for verification of nitrogen components and organic carbon, but also for calibration and validation of microbial community dynamics (Ni et al., 2012). However, low-quality long-term experimental data can hamper the verification of biokinetic models on a statistical basis (Kovárová-Kovar and Egli, 1998). To overcome the claimed challenges, this study investigates the modeling of simultaneous partial nitrification, anammox and denitrification (SNAD) in a full-scale biofilm reactor assisted by granular activated carbon (GAC) particles under anoxic conditions.

The main aim of the work is to develop and evaluate a comprehensive and novel mechanistic model including SMP and EPS definitions to explain the relative abundance of independent microbial groups and to further describe the dynamics of microbial groups in the biofilm. Specific aims include:

(1) Optimization of an inexpensive cytogenetic technique to estimate the active microbial fractions in the granules. The observed values from optimized technique can be later used to verify the results of simulation of microbial population dynamics.

(2) Development and application a robust protocol to rank the sensitive parameters, calibrate, validate, and check the accuracy of the model performance for carbon, nutrients and microbial relative abundances.

(3) Comparison and analyze two different formats of biofilm reactor model simulations with and without EPS and SMP definition.

(4) Prediction of the time-dependent concentration of SMP and EPS.

At the end, the main research questions to be answered is to evaluate whether and when it is necessary to define and apply such a complicated biofilm modeling framework using EPS matrix definition for simulation of biological wastewater treatment plants? And how to put that informative model into practice in the wastewater treatment industry?

## 4.2 Material and methods

### 4.2.1 Operation of biofilm reactor

The previously explained full-scale anammox plant for landfill leachate treatment (explained in chapter 1) operated by LAMBDA GmbH (Herten, Germany) consisting of three main stages, including biological treatment with activated sludge, ultrafiltration, and activated carbon biofilm process (Azari et al., 2017a) (Fig. 27). The entire processes are very stable to sudden variations in flow and nitrogen concentration, and it can convert nearly 97 % of the ammonium to nitrogen gas. In this study, the last part of the plant before the outflow containing GAC-assisted biofilm reactors (essential for final polishing of remaining amount of ammonium, nitrite, and nitrate after the activated sludge process) was modeled. The anammox organisms grow on the surface of activated carbon with silica gel as the adsorbent medium. The GAC biofilm also comprises nitrifying and heterotrophic denitrifying bacteria (Ke et al., 2015). Daily concentrations of $NH_4$-N, $NO_2$-N and $NO_3$-N were recorded photometrically using German standards (Hach, Germany). Total organic nitrogen (TON) is negligible in the influent and effluent of the biofilm reactor and pH value is controlled between 7 and 7.4. Total inorganic nitrogen (TIN) was calculated as the sum of $NH_4$-N, $NO_2$-N and $NO_3$-N.

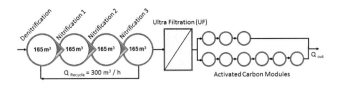

Figure 27: Illustration of simulated anammox plant with combination of activated sludge (biological treatment) and activated carbon biofilm system (biophysical treatment) in the plant.

### 4.2.2 Quantitative fluorescence *in-situ* hybridization (qFISH) and image analysis

Series of experiment was designed to develop a protocol adapted from for qFISH, *in-situ* microscopy and image analysis for quantification of relative abundance of dominant microorganisms (see Supplementary Materials A). The biomass was frequently taken from the steady-state operation of the GAC biofilm reactor in five monthly timeframes. Three group of bacteria were targeted including anammox (AMX), ammonia-oxidizing bacteria (AOB) and nitrite-oxidizing bacteria (NOB). Totally, five specific probes mixed with a general probe were chosen (Table 8). Probe Amx820 and Sca1309 target most anammox bacteria, Nso190 targets most AOB, NIT3 and Ntspa662 target most NOB, and EUB338 is a general probe to target most domain bacteria (more than 90 %). 20 qFISH images from randomized field of observations (FOVs) were obtained by an epifluorescence microscope (Axio imager 2, Carl Zeiss AG, Jena, Germany). The relative abundances of AMX, AOB and

NOB were estimated by calculating the percentage of their respective areas taken up by fluorescence targeted cells from the specific probes compared to the total area taken up by theEUB338 general probe using ImageJ and daime software (Daims et al., 2006; Almstrand et al., 2014).The relative abundance using qFISH, will be further used for model evaluation.

Table 10: 16S rRNA-targeted oligonucleotide gene probes used for qFISH

| Probe | Specificity | Sequence (5'-3') | Formamide [%] | Dye | References |
|---|---|---|---|---|---|
| EUB 338 | Most of domain bacteria | GCT GCC TCC CGT AGG AGT | 0 – 55 | FITC | (Zarda et al., 1991) |
| Amx 820 | AMX: *Candidatus* Brocadia anammoxidans and *Candidatus* Kuenenia stuttgartiensis | AAA ACC CCT CTA CTT AGT GCC C | 40 | Cy3 | (Kartal et al., 2006) |
| Sca 1309 | AMX: *Candidatus* Scalindua sp. | TGG AGG CGA ATT TCA GCC TCC | 5 | Cy3 | (Schmid et al., 2003) |
| NIT 3 | NOB: *Nitrobacter* sp. | CCT GTG CTC CAT GCT CCG | 40 | Cy3 | (Mobarry et al., 1997) |
| Ntspa 662 | NOB: *Nitrospira* sp. | GGA ATT CCG CTC TCC TCT | 35 | Cy3 | (Daims et al., 2000) |
| Nso 190 | AOB: Betaproteobacterial ammonia-oxidizing bacteria | CGA TCC CCT GCT TTT CTC C | 55 | Cy3 | (Mobarry et al., 1997) |

### 4.2.3 Model definition and evaluation

The previously discussed model in the last chapter which was verified for short-term (Azari et al., 2017c) based on Activated Sludge Model No. 1 (ASM1) on the biofilm modeling platform in AQUASIM 2.1 (EAWAG, Switzerland), was modified for enhanced nitrogen elimination including four main microbial groups i.e. AOB, NOB, AMX and heterotrophic denitrifiers (Table 2) according to the Monod kinetics and the revised stoichiometric matrix of the anammox process (Lotti et al, 2014). An additional definition of soluble microbial products (SMP) and EPS was added to the model structure to compare the model performance with and without EPS and SMP kinetics. This structure can model SMP production and uptake, EPS production and EPS hydrolysis rates according to Liu et al. (2016). SMPs were classified into two groups, utilization-associate products (UAP) from the natural product of bacterial growth and biomass-associate products (BAP) from the hydrolysis process of the active biomass (Laspidou & Rittmann., 2002). For a better insight, a schematic and artificial demonstration of placement and distribution of distinct groups of bacteria considered in the model with EPS for an assumed day of the simulation showed in Fig. 28 The inert ($X_i$) and slowly biodegradable ($X_s$) located in the inner layer of the granular biofilm (grey layer), anammox ($X_{AN}$) and heterotrophs ($X_H$) were the main active biomass distributed in the anoxic zone of the biofilm, while AOB ($X_{AOB}$) and NOB ($X_{NOB}$) coexisted at the outer layer. The EPS ($X_{EPS}$) and other living biomass (Xothers) proposed to distribute along the thickness of the biofilm (Table 11). As previously discussed in the introduction, it is essential to establish a robust protocol for model evaluation. Therefore, this work formulated a protocol adapted from Zhu et al., (2016) for experiment design, data collection, model development, sensitivity analysis, calibration, validation, and uncertainty analysis (Fig. 29). The new contributions to this model evaluation algorithm are (1) adding the experimental design procedure for qFISH and (2) defining statistical criteria for model calibration and validation for TIN and the relative abundance of microbial consortia. Fig. 30a and 30b illustrate how two model structures with and without EPS kinetics were developed. Since GAC has the potential to adsorb COD and $NH_4$-N, a simple adsorption isotherm equation ($K_d$ approach) as a linear form of the Freundlich equation was also added to both models to define adsorption kinetics (Ranjbar and Jalali, 2015). Bioprocess kinetic rates equations including those for EPS, UAP and BAP are presented in Table 10. The stoichiometric matrix and definition of biofilm matrix parameters are respectively provided in Table D1 and D2, respectively, of the Supplementary Materials.

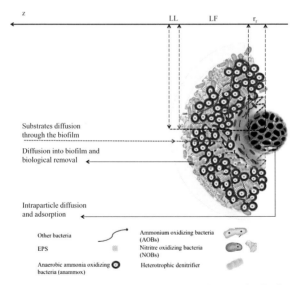

Figure 28: An artificial schematic view of the model showing the distribution of anammox, AOB, NOB, heterotrophic denitrifiers, other bacteria and EPS as well as the transport,

removal and flux of the substrate through a GAC covered by microbial biofilm for the simulation results at the day 50. Inert fraction (grey layer) is assumed as a fraction of granular biomass thickness ($L_f$) and GAC (black porous sphere) is assumed as the support particle with the constant radius ($r_r$). LL is the boundary layer thickness.

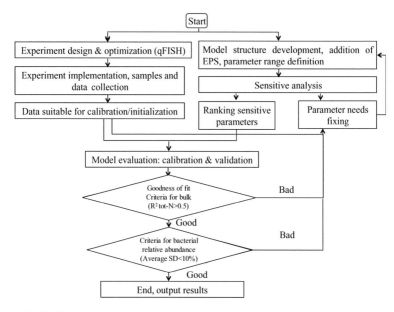

Figure 29: Model development and evaluation protocol applied in this study.

Table 11: Definition of components in the models

| Component | Definition | Unit |
|---|---|---|
| **Dissolved components** | | |
| $S_{BAP}{}^*$ | Biomass associated products | gCOD m$^{-3}$ |
| $S_{UAP}{}^*$ | Utilization-associated products | gCOD m$^{-3}$ |
| $S_S$ or $S_{COD}$ | Readily degradable organic substrate | gCOD m$^{-3}$ |
| $S_{NH4}$ | Ammonium nitrogen | gN m$^{-3}$ |
| $S_{NO2}$ | Nitrite nitrogen | gN m$^{-3}$ |
| $S_{NO3}$ | Nitrate nitrogen | gN m$^{-3}$ |
| $S_{N2}$ | Dissolved nitrogen gas | gN m$^{-3}$ |
| **Model particulate components** | | |
| $X_{AN}$ | Anaerobic ammonium-oxidizing bacteria | gCOD m$^{-3}$ |
| $X_{AOB}$ | Ammonium-oxidizing bacteria | gCOD m$^{-3}$ |
| $X_{NOB}$ | Nitrite-oxidizing bacteria | gCOD m$^{-3}$ |
| $X_{EPS}{}^*$ | Extracellular polymeric substances | gCOD m$^{-3}$ |

| $X_H$ | Heterotrophic bacteria | gCOD m$^{-3}$ |
|---|---|---|
| $X_S$ | Slowly degradable organic substrate | gCOD m$^{-3}$ |
| $X_I$ | Inert, non-biodegradable organics | gCOD m$^{-3}$ |
| $X_{others}$ | Other bacteria, not contributing to carbon and nitrogen removal processes | gCOD m$^{-3}$ |

*\* Components were only available in model with EPS*

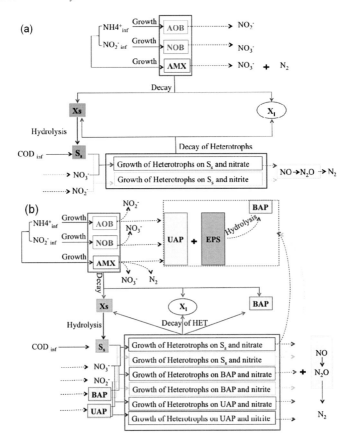

Figure 30: The model structure without (a) and with (b) the EPS kinetics and dynamics of SMP release and uptake a for a biofilm system in anoxic conditions containing four groups of bacteria: AMX, AOB, NOB and heterotrophic denitrifiers (UAP: utilization-associated products and BAP: biomass-associated products).

Table 12: Bioprocess kinetic rate equations for models

| Number | Process | Kinetics rates equation |
|---|---|---|
| 1 | Hydrolysis of slowly biodegradable substrates | $K_H \dfrac{\left(X_S / X_H\right)}{\left(K_X + X_S / X_H\right)} X_H$ |
| 2* | Hydrolysis of EPS | $K_{H\_EPS} \, X_{EPS}$ |
| 3 | Growth of anammox | $\mu_{ANmax} \dfrac{K_{O_2}^{AN}}{K_{O_2}^{AN} + S_{O_2}} \dfrac{S_{NH_4}}{K_{NH_4}^{AN} + S_{NH_4}} \dfrac{S_{NO_2}}{K_{NO_2}^{AN} + S_{NO_2}} X_{AN}$ |
| 4 | Growth of AOB with two steps nitrification | $\mu_{AOBmax} \dfrac{S_{O_2}}{K_{O_2}^{AOB} + S_{O_2}} \dfrac{S_{NH_4}}{K_{NH_4}^{AOB} + S_{NH_4}} X_{AOB}$ |
| 5 | Growth of NOB with two steps nitrification | $\mu_{NOBmax} \dfrac{S_{O_2}}{K_{O_2}^{NOB} + S_{O_2}} \dfrac{S_{NH_4}}{K_{NH_4}^{NOB} + S_{NH_4}} \dfrac{S_{NO_2}}{K_{NO_2}^{NOB} + S_{NO_2}} X_{NOB}$ |
| 6 | Decay of anammox | $b_{AN} \, X_{AN}$ |
| 7 | Decay of AOB | $b_{AOB} \, X_{AOB}$ |
| 8 | Decay of NOB | $b_{NOB} \, X_{NOB}$ |
| 9 | Anoxic growth of $X_H$ on nitrite and $S_S$ | $\mu_{Hmax} \, \eta_{NOX} \dfrac{K_{O_2}^{H}}{K_{O_2}^{H} + S_{O_2}} \dfrac{S_{NO_2}}{K_{NO_2}^{H} + S_{NO_2}} \dfrac{S_S}{K_S^{H} + S_S} X_H$ |
| 10 | Anoxic growth of $X_H$ on nitrate and $S_S$ | $\mu_{Hmax} \, \eta_{NOX} \dfrac{K_{O_2}^{H}}{K_{O_2}^{H} + S_{O_2}} \dfrac{S_{NO_3}}{K_{NO_3}^{H} + S_{NO_3}} \dfrac{S_S}{K_S^{H} + S_S} X_H$ |
| 11* | Anoxic growth of $X_H$ on nitrite and BAP | $\mu_{Hmax\_BAP} \, \eta_{NOX} \dfrac{K_{O_2}^{H}}{K_{O_2}^{H} + S_{O_2}} \dfrac{S_{NO_2}}{K_{NO_2}^{H} + S_{NO_2}} \dfrac{S_{BAP}}{K_{BAP}^{H} + S_{BAP}} X_H$ |
| 12* | Anoxic growth of $X_H$ on nitrate and BAP | $\mu_{Hmax\_BAP} \, \eta_{NOX} \dfrac{K_{O_2}^{H}}{K_{O_2}^{H} + S_{O_2}} \dfrac{S_{NO_3}}{K_{NO_3}^{H} + S_{NO_3}} \dfrac{S_{BAP}}{K_{BAP}^{H} + S_{BAP}} X_H$ |
| 13* | Anoxic growth of $X_H$ on nitrite and UAP | $\mu_{Hmax\_UAP} \, \eta_{NOX} \dfrac{K_{O_2}^{H}}{K_{O_2}^{H} + S_{O_2}} \dfrac{S_{NO_2}}{K_{NO_2}^{H} + S_{NO_2}} \dfrac{S_{UAP}}{K_{UAP}^{H} + S_{UAP}} X_H$ |
| 14* | Anoxic growth of $X_H$ on nitrate and UAP | $\mu_{Hmax\_UAP} \, \eta_{NOX} \dfrac{K_{O_2}^{H}}{K_{O_2}^{H} + S_{O_2}} \dfrac{S_{NO_3}}{K_{NO_3}^{H} + S_{NO_3}} \dfrac{S_{UAP}}{K_{UAP}^{H} + S_{UAP}} X_H$ |
| 15 | Decay of $X_H$ | $b_H \, X_H$ |

* Process is only available in model with EPS

## 4.3 Results and discussion

### 4.3.1 Simulation of nitrogen removal

The general behaviour of the biofilm reactor's influent is characterized by significant fluctuations in $NH_4$-N and COD with a COD:TN ratio of 4 to 20. The one-year performance of the anammox biofilm reactor was calibrated for two models using measured data of $NH_4$-N, $NO_2$-N and $NO_3$-N. The regression analysis between the simulated and measured concentration of TIN and $NH_4$-N implies a higher $R^2$ value for the model with EPS and SMP. For the model with EPS and SMP, $R^2$ values for TIN and $NH_4$-N are equal to 0.65 and 0.66 respectively while $R^2$ values of 0.64 and 0.61 were calculated for the model without EPS kinetics. Calibrated values for model parameters and regression analysis graphs for TIN and

NH$_4$-N are provided in Table D3 and D4 and Fig. D5 of the Supplementary Materials. After the model calibration, validation results of both models for TIN over 120 days of simulation showed that the model with EPS and without EPS predict the fate of remaining nitrogen in the effluent with a similar accuracy. Although the model with EPS undermined some measured data in the middle of validation period, the model without EPS overestimated some data points in the beginning of validation period (Fig. D6 of Supplementary Materials). Nevertheless, after regression analysis both models gave the identical R$^2$ value of 0.65 for the validation period.

### 4.3.2 Model evaluation for relative abundance of microbial consortia

The relative abundances of AMX, AOB and NOB were visualized and estimated by qFISH. As shown in Fig. 31 averaged relative abundance of anammox bacteria as dominant species to the total domain bacteria over one year was 52.2 % ± 5.4 of total microbial groups while the simulated results give the averaged relative abundance of 45.8 % for the model without EPS (Fig. 32a) and 53.7 % for the model with EPS (Fig. 32b). For AOB, the average of estimated relative abundance of bacteria to the total domain bacteria throughout the year was 5.7 % of total microbial groups while the modeled results give a relative abundance of 15.8 % for the model without EPS (Fig. 32a) and 8.5 % for the model with EPS (Fig. 31b). The NOB was the least dominant bacterial group, at only 2.6 ± 3.2 % from qFISH results (Fig. 31). Simulation of NOB relative abundance was in good agreement with the observations: the model with and without EPS predicted 2.6 ± 1.8 % and 3.3 ± 2.8 % respectively (Fig. 32a, b). Comparing the simulated relative abundance with the observed values for AMX, AOB and NOB, it could be seen that statistically the model with EPS fits better than model without EPS, in which averaged relative abundance was closer to the measured data and the trends and variations were better predicted (Fig. 32b). Moreover, throughout the simulation period, the model without EPS overestimated the AOB relative abundance while the relative abundance of heterotrophs decreased continuously to less than 10 %. This could have been due to the underestimate of heterotrophic growth rates due to neglecting heterotrophic synthesis from the utilization of UAP and BAP as extra organic carbon sources.

With regards to the kinetics of BAP and UAP formation and uptake, UAP and BAP have distinct kinetics parameters, with UAP more readily biodegradable for heterotrophs. Considering the set parameters in our model for UAP and BAP, the specific growth rates, and the yield coefficient of readily soluble COD (S$_s$) is higher than UAP and for UAP is higher than BAP. On the other hand, the affinity constant for BAP is higher than UAP and for UAP is higher than readily soluble COD ($S_s$) (See Table D4). These reasons will lead to the favorable consumption of UAP than BAP by heterotrophs. This finding follows earlier theoretical hypotheses to the most recent findings claiming that UAP is more important than biomass-associated SMP (BAP) (Namkung and Rittmann, 1986; Liu et al, 2016)

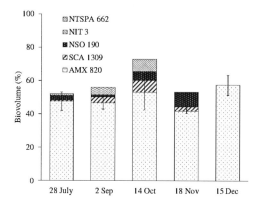

Figure 31: Relative abundance of anammox (AMX820 and SCA1309), AOB (NSO190) and NOB (NTSP662 & NIT3) in biofilm samples from July, September, October, November and December in 2016 estimated from qFISH experiments.

### 4.3.3    Correlation analysis of nitrogen removal efficiency and biomass fractionations

The model predicts total biofilm fractionations in terms of percentage of active (microbial) groups e.g. heterotrophic denitrifiers, anammox, AOB and NOB as well as inactive (non-microbial) groups e.g. EPS, inert, and slowly biodegradable matter (Fig. 33). To link the fractions of microbial and non-microbial groups in the biofilm with system performance, the observed and modeled nitrogen removal efficiency (NRE) (average of 10 days' values) are sketched. Inert matter contributed 37.5-44 % of biomass and EPS fraction varied 6-10 % of biofilm matrix. After 15 days of simulation period, the fraction of autotrophs and EPS was stable, but the fraction of heterotrophs decreased, and the fraction of inert matter rose while the NRE fluctuated between 37 % and 82 % (Fig.33a and 33b, black connected line). The highest removal efficiency of the system was found between days 90 and 250, which can be predicted by models as well. Nevertheless, no specific link can be found to correlate the system performance in terms of efficiency and stability with fractions of microbial and non-microbial groups in the biofilm.

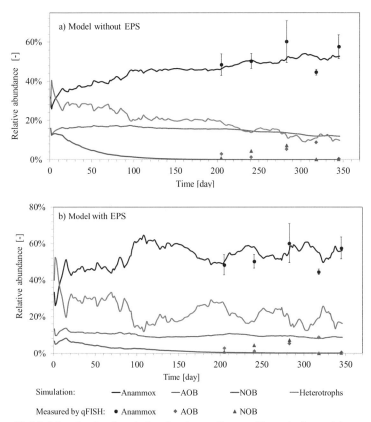

Simulation: —Anammox —AOB —NOB —Heterotrophs

Measured by qFISH: ● Anammox ♦ AOB ▲ NOB

Figure 32: Model evaluation for relative abundance of targeted bacteria for models

### 4.3.4 Correlation analysis of the nitrogen surface loading and biomass fractionations

The nitrogen surface loading (NSL) was defined as an average of TIN loaded on a square meter of biofilm active surface. In the first 150 days of the simulation, the nitrogen surface loading was higher, estimated to be at 59-105 gN $m^{-2}$ $d^{-1}$ for the model with EPS and from 44.1-85.0 gN $m^{-2}$ $d^{-1}$ for the model without EPS (Fig. 33, yellow connected line). But after the day 150, NSL declines to less than 20 gN $m^{-2}$ $d^{-1}$ due to an increase in biofilm thickness for both models. Consequently, the concentration of heterotrophs, autotrophs (AMX, AOB and NOB) and EPS slightly decreased, when the biofilm thickness increased from 0.62 to 1.2 mm for the model without EPS and from 0.57 to 1.01 mm for the model with EPS. Nevertheless, over the one-year simulation period, the fractions of autotrophs and EPS were stable, but the fraction of heterotrophs decreased significantly which is correlated with reduction in nitrogen surface loading of the biofilm. Besides, the fraction of inert matter in the biofilm correlates inversely with NSL values.

Total findings are compatible with a study from Ni et al (2012) demonstrated how the NSL impacts on the biomass fractions of anammox biofilm and it indicated that during a long-term simulation, the NSL increase will lead to a raise in the fractions of heterotrophs, EPS, and anammox and a reduction in the fraction of inert. After a while, these fractions will be levelled off even though the nitrogen loading rate is increasing (Ni et al., 2012).

### 4.3.5 Dynamic model prediction of SMP and EPS

The model with EPS adapted from the unified theory of Laspidou and Rittmann, (2002a,b), which combines the production and consumption of SMP with the formation and degradation of EPS within biofilm system. Using this modeling approach, EPS and SMP products in the granular biofilm reactor were quantitatively analyzed and the results are shown in Fig. 34.

The SMP products from the biofilm include BAP (biomass-associated products), which are products of biomass decay and EPS hydrolysis, and UAP (utilization-associated products), which are formed during the growth of microorganisms. The concentration of UAP in the effluent was lower and quite stable at $2.68 \pm 0.16$ mg COD $L^{-1}$, while the predicted BAP concentration was fluctuated in the range between 2.11 and 12.94 mg COD $L^{-1}$ (Fig. 34). It means that the BAP was the main contributor of SMP products in the effluent. Ni et al (2012) modeled SMP dynamic in the anammox biofilm reactor was modeled for 1,000 days. The results indicated that total SMP (UAP+BAP) concentration could reach to 17 mg COD $L^{-1}$ (Ni et al., 2012). Another work from Liu et al (2016) focusing on the growth of heterotrophs on SMP products in the anammox biofilm showed a minor portion of UAP contributed for SMP in the effluent, which SMP concentration predicted from 6.0 to 6.25 mg COD $L^{-1}$ but the concentration of UAP was 3 order of magnitude lower than total SMP (Liu et al., 2016).

Regarding the EPS concentration, as it was mentioned in the last section there is a reduction in the EPS concentration and its relative abundance. This is a direct consequence of reduction in the EPS production due to the reduction in the nitrogen and organic carbon surface loading during the development of biofilm area. Because the yield of the EPS on heterotrophs is higher than anammox (since anammox has relatively slower growth rate) the dependency of EPS growth is highly attributed to the heterotrophic growth rate. Hence, heterotrophic decay will highly contribute on loss of EPS concentration.

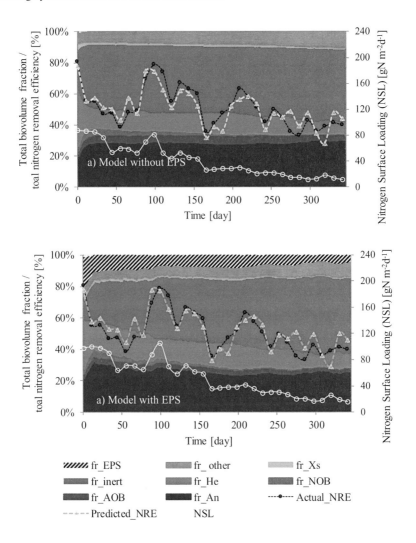

Figure 33: Correlative assessment of biomass fractions and the system performance as TIN removal efficiency (NRE) and nitrogen surface loading (NSL) for the model without EPS (a) and with EPS (b).

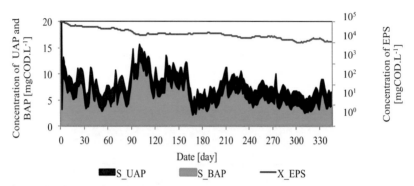

Figure 34: The model prediction of soluble microbial products including BAP and UAP in the effluent and the concentration of the EPS in the biofilm matrix

With regards to the kinetics of BAP and UAP formation and uptake due to the growth of biomass, UAP concentrations are much lower than BAP, indicating that the heterotrophs prefer to uptake UAP than BAP. This leads to the necessity to investigate the relative importance of UAP and BAP for the heterotrophic growth in anoxic anammox biofilms:

1) UAP and BAP have distinct kinetics parameters, with UAP more readily biodegradable for heterotrophs. Considering the calibrated parameters in the model for UAP and BAP, the specific growth rates and the yield coefficient of readily soluble substrate ($S_s$) is higher than UAP and for UAP is higher than BAP. On the other hand, the affinity constant for BAP is higher than UAP and for UAP is higher than readily soluble substrate ($S_s$) (see Table 9). These reasons will lead to this assumption that there is a favorable consumption of UAP than BAP by heterotrophs. This finding follows earlier theoretical hypotheses to the most recent model-based findings claiming that UAP is more important than biomass-associated SMP (BAP) (Namkung and Rittmann, 1986; Liu et al, 2016)

2) Due to the increase in the inert matter concentration in the consortia because of a predicted heterotrophic decay (due to the reduction in the nitrogen surface loading) and since heterotrophic denitrifiers demonstrated a higher decay rate than anammox, the amount of released BAP will be enormously increased during their decay. One should notice that there is a key parameter called $f_{BAP}$, which appoints the converted fraction of BAP during the decay of all microorganisms. Accordingly, a higher $f_{BAP}$ leads to a striking release of BAP as a product of decay. As a result, the BAP accounts for most of the SMP concentration rather than the UAP.

3) Besides, it is important to mention that the production and biodegradation of EPS and SMPs are also correlated with shifts in the solid retention time which are not considered directly in the kinetics of this model independently and hence, it might impact on the behavior and the prediction bias (Jarusutthirak and Amy, 2006, Chen et al., 2012).

## 4.4   Conclusion

This kinetic study proposed a novel model structure for biological nitrogen removal by including EPS, UAP and BAP components and new process rates, which made ASM1 more complete. The model is verified for GAC biofilm reactors and results elucidated that the model can simulate inorganic nitrogen components in the bulk liquid and the relative abundance of targeted bacterial groups with enough accuracy. The key findings are:

The model was improved after the addition of EPS and SMP kinetics.

- The model proved that the microbial and non-microbial composition of a stable biofilm reactor is not correlated with the nitrogen removal efficiency, but it might be more influenced by any change in operation conditions such as flow rate, hydraulic retention time, dissolved oxygen, temperature, salinity, and pH level.

- The model without EPS and SMP kinetics overestimated the abundance of AOB.

- The model without EPS underestimated heterotrophic growth compared with the model with EPS because the model with EPS and SMP incorporates UAP and BAP for the support of heterotrophic growth.

- UAP is more readily biodegradable for heterotrophs than BAP.

- The real-time production of EPS and SMP can be simulated but further works are required to evaluate and validate prediction of EPS and SMP.

This work suggests that the growth and existence of heterotrophs in anammox biofilm systems increases due to incorporating the production of SMP so that SMP supports the growth of heterotrophic denitrifiers. Hence, application of the model with EPS is an advantage when we want a better insight on microbial consortia. With a reliable prediction of SMP release, we can better calculate and control soluble COD and inorganic nitrogen concentrations   in   the   effluent   of   wastewater   treatment   plants.

# 5 The effect of regulation of aeration and temperature on nitrogen removal and microbial community structure in a hybrid sequencing batch reactor

## 5.1 Introduction

SNAD process is a complex biological process where different populations with opposed environmental requirements coexist. It combines partial nitritation process, anammox process and denitrification process simultaneously in one single reactor under oxygen limitation. Thus, the conditions are micro-aerobic for partial nitritation and anoxic for anammox and denitrification. SNAD process has various advantages compared to two-stage systems and other conventional biological treatments since it has lower capital costs, no external carbon sources, lower sludge productions and lower energy & oxygen requirements. A key factor for the development of the SNAD process is a better understanding of both the biological processes involved in the nitrogen removal and the numerous microbial interactions using microbiological and modeling tools (Langone, 2013). Microbial community play a very important role in controlling SNAD. Ammonia oxidizing bacteria (AOB), nitrogen oxidizing bacteria (NOB) and anammox bacteria or anaerobic ammonia oxidizing bacteria (AnAOB) are the most distinguished bacteria for wastewater treatment.

Yet, biofilm reactors are the most frequently used process to retrofit municipal WWTPs for enhanced and energy efficient biological nitrogen removal including one stage SNAD process (Liu et al., 2017). Biofilms are microbial communities characterized by their adhesion to any solid support particle or surface and the production of a dense matrix of extracellular polymeric substances (EPS), consisting of polysaccharides, proteins, DNA and lipids, which surround the microorganisms lending structural integrity and a unique biochemical profile to the biofilm and enhances ability of microbial persistence in each environment resulting in keeping solid retention time high enough (Coughlan et al., 2017). Various anoxic and aerobic granular sludge can be also considered as a special case of biofilm, where self-immobilized bacterial cells form strong, dense, and well-settling aggregates (Show et al., 2012). Processes containing both suspended biomass and biofilm, usually referred as hybrid systems. For instance, IFAS process, is created by introducing plastic elements as biofilm carrier media (like MBBRs) into a conventional activated sludge reactor (Regmi et al, 2016, Zhang et al., 2015). So far, the most representative biofilm-based reactors for biological nitrogen removal include moving bed biofilm reactor (MBBR), and integrated fixed-film activated sludge system (IFAS) (Liu et al., 2017, Zhang et al, 2015).

Besides, granular activated carbon (GAC) was also used in an anammox process with the aim of evaluating its use as a support particle to enhance biofilm formation (Wenjie et al., 2015; Azari et al, 2017).

In the nitrification process, the rise of temperature creates two opposite effects including the increase of free ammonia inhibition and increased activity of the organisms. For partial nitritation process, the optimal temperature is between 35°-45 °C (Van Hulle et al., 2007). Additionally, anammox activity can be expected at $10° - 35°$ C. The long-term experiments were performed at temperature between 15-30°C and the activity was lost at 15°C (Dosta et al., 2008) and the system couldn't remove all the $NO_2^-$-N applied in the reactor, which causing inhibition for anammox activity later. It was strongly recommended to have slow adaptation of anammox process at low temperature. Although there was also $NO_2^-$-N accumulation at temperatures of 12°-15 °C, the anammox activity recovered quickly even while the temperature was further decreased. Low temperatures in winter might also help to suppress NOB, as they were affected most by temperatures below 13°C and showing no recovery. Anammox bacteria, on the other hand, were able to shift their temperature optimum and recover at temperatures below 13°C.

In terms of effect of aeration regime to nitrogen removal, long sludge retention time (SRT) might be employing intermittent aeration to accomplish SNAD in a single reactor. Aeration pattern is an alternative for controlling partial nitrition (Hidaka et al., 2002). The duration of aeration is contrarily related to the extent of partial nitrition due to the conversion from partial to complete nitrition at long aeration periods (Turk & Mavinic, 1989). Thus, an intermittent or discontinuously aeration can be an option for effective partial nitrition (Lee et al., 1999). Intermittent aeration systems typically are operated with SRT values in the range of 18 – 40 days and hydraulic retention time (HRT) in excess 16 hours. During the aeration-off period, the reactor operates as an anoxic reactor and the mixer will be switched on (mixing phase) and nitrate will be used as the electron acceptor. The time for the anoxic and aerobic periods is important in determining the performance of the treatment system. Thus, it could be adjusted manually as part of the system operation to optimize the process performance. Reported plant performance data for intermittent aeration process indicate effluent $NO_3$-N concentrations range from 3-4.8 mg L$^{-1}$. The relatively long SRT values used provide sufficient dilution to minimize the effluent $NH_4^+$-N concentrations during the OFF period. A sufficiently long SRT is also needed to provide enough nitrification capacity to allow the aeration system to be operated intermittently (Tchobanoglous et al., 2003).

Objectives of this research are (i) to operate two SBRs one as IFAS and another with combination of biofilm on activated carbon and activated sludge flocs and to compare which reactor has a better potential to reach a faster start-up and a higher efficiency during short time. (ii) to analyse the microbial community diversity in both systems using qFISH and high-throughput 16S rRNA gene amplicon sequencing and linking it to the system performance. (iii) to evaluate effect of reduced temperature gradient on nitrogen removal performance and specific activities. The idea is to reach a sustainable and stable growth of anammox planctomycetes majorly attached to biofilm carriers or in the granules and AOB majorly lying in as flocculent sludge and planktonic cells in activated sludge using a hybrid IFAS system in a SBR.

## 5.2 Material and Methods

### 5.2.1 Experimental setup and operational conditions

Two identical lab-scale hybrid sequencing batch reactors (SBRs) with a working volume of 5 L were used to reach the purpose of enhanced nitrogen removal with SNAD. Reactor 1 was based on the integration of biofilm carrier technology with conventional activated sludge (IFAS - integrated fixed film activated sludge). Reactor 2 was operated with mixture of biofilm formed on granular activated carbon (GAC) as a microbial carrier and activated sludge. The inoculating biomass of activated sludge for both reactors were obtained from activated sludge part of a leachate treatment plant in Zentraldeponie Emscherbruch (ZDE) operated by AGR group (Herten, Germany) which was previously explained in Chapter 2 and 4 (Azari et al., 2017a). The biofilms for the first reactor were K1 Kaldnes carrier (AnoxKaldnes, Lund, Sweden) from an operating municipal waste water treatment plant in Hattingen (Ruhrverband, Essen, Germany) with filling ration of 15 %. Besides, fresh K3 carriers (AnoxKaldnes, Lund, Sweden) without any inoculated biomass with filling ratio of 25 % were added. Granular sludge (biofilm formed on GACs) for second reactor was obtained from activated carbon biofilm part of the same plant (Azari et al., 2017a). The influent was a synthetic wastewater (the composition is given in Table 14) and operational parameters were monitored by a PLC system.

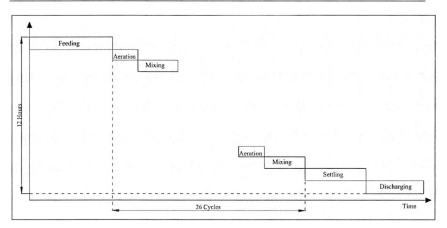

Figure 35: 12 hours cycle of IFAS-SBR


Each cycle of the SBR system consisted of filling, intermittent aeration and anoxic mixing, settling, decanting, and idling. There were many repeated cycles of the intermittent aeration and anoxic mixing (Fig. 51). There were 2 cycles (AM and PM cycle) in total for one operation per day. Feeding, aeration, mixing and discharging time were controlled by IKS Program (Aquasoft 'Aquastar', iks ComputerSysteme GmbH, Germany) as well as the pH, dissolved oxygen concentration (DO), redox potential (Eh) and temperature through online sensors. The pH-sensors were divided into pH-Base and pH-Acid sensors along with each solution, an 0.02 M NaOH and 0.02 M HCl. The pH was controlled with 6.5 to 8 limits. If the pH-limit goes up, HCl solution will be pumped into the reactor and vice versa. The feeding and discharging were done by peristaltic pumps volume was kept between 1.5 – 1.7 L and the hydraulic retention time (HRT) was controlled approximately at 36 h which seems to be an ideal for the operation of deammonification process in SBRs (Anjali & Sabumon, 2017).

A refrigeration circulator (Kältethermostat, witeg Labortechnik GmbH, Germany) was used to aim a stabile decline of temperature. To find the optimum of oxygen concentration and optimum sequenced aeration regime, the aeration was operated with varied duration and controlled with a manual valve (iNFLUX, Influx Measurements Ltd, UK) at between 0.1 to 0.5 L min$^{-1}$. The SBR was mixed using a magnetic stirrer (IKAMAG RCT, IKA-Werke GmbH & Co. KG, Germany) with 300 – 500 rpm as to avoid erosion on bottom of the reactor. Mixing phase was lasted for 13 – 20 minutes for each cycle in the interest of a homogeneous condition thereby was also influencing the growth of anammox bacteria. The gas effluent was being flowed to a gas outlet, as shown in Fig. 36. The recorded data was read and saved once every two working days, as in Monday, Wednesday, and Friday. To have a correspondence data, nitrogen components and COD measurement was conducted also in those days. The water level on the reactor was specifically necessary to be manually read and written in report book because the unavailability of water level sensor.

### 5.2.2 Bioaugmentation scenarios

Also, regular bioaugmentation using various granular sludge from off-line full-scale PN-A plants with acclimated biomass grown in reactor with long SRT will be used to augment our lab-scale plant. Later, granular sludge from ZDA, a landfill leachate treatment plant located in a municipal waste disposal site in Alsdorf, Aachen was added since day 23 until day 86 of operation days. From day 98 until day 126 more sludge from ZDE plant was also used for bioaugmentation purposes. Lastly due to failures in ZDE and ZDA plants, on the day 166, a new biomass from reject water treatment of municipal wastewater plant in Kamen (Emschergenossenschaft) in the form of granular sludge was used also for bioaugmentation. The amount of sludge for bioaugmentation were assorted between 150-300 mL, which based on actual condition of the reactor and VSS results (Table 13). All biomass was collected fresh, transported on the same collection day, and kept in fridge with $4°$ C.

Table 13: MLSS and MLVSS of seeding sludge used for bioaugmentation of hybrid IFAS-SBR.

| Date | | MLSS | MLVSS | Sludge |
|---|---|---|---|---|
| Tested | Added since | g L$^{-1}$ | g L$^{-1}$ | |
| 04-Sep-17 | 07-Jul-17 | 5.8 | 5.4 | ZDA |
| 15-Sep-17 | 20-Sep-17 | 11.8 | 7.1 | ZDE |
| 18-Dec-17 | 27- Nov-17 | 5.6 | 3.9 | Kamen |

### 5.2.3 Mineral solutions

The chemicals for the synthetic water are stated in Table 14. Synthetic water was made for 50 L influent volume, although sometimes the influent was made only to 25 L volume to avoid the unwanted process, such as nitritation in the feed tank. Hence, the feed tank will be rinsed, and new synthetic water will be added in once a week. On day 22, 06th of July 2017, $(NH_4)_2SO_4$ was reduced to 10 % while $NaHCO_3$ increased to 10 % for stabilization of the initial phase. The gradual reduction of $(NH_4)_2SO_4$ was conducted to know whether the amount of it influenced the nitrogen removal efficiency.

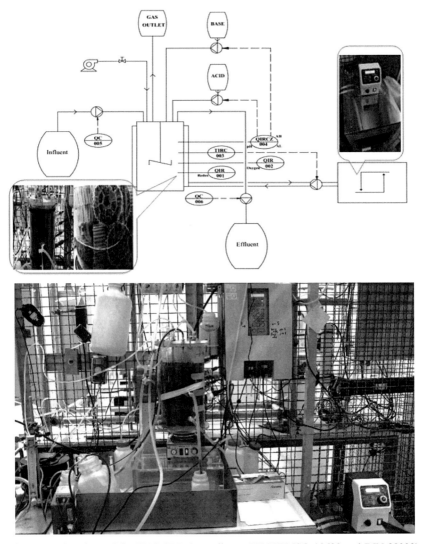

Figure 36. Flowchart of the IFAS-SBR (according to DIN EN ISO 10628 and DIN 28000). IFAS-SBR on the day 180.

Table 14: List of chemical substances for synthetic water

| Chemical | Formula | Concentration (mg L$^{-1}$) |
|---|---|---|
| Sodium hydrogen carbonate | $NaHCO_3$ | 1,100 |
| Potassium permanganate | $KH_2PO_4$ | 27 |
| Magnesium sulphate hydrate | $MgSO_4.7H_2O$ | 300 |
| Calcium chloride dihydrate | $CaCl_2.2H_2O$ | 300 |
| Iron (III) sulphate heptahydrate | $FeSO_4.7H_2O$ | 12.1 |
| EDTA disodium salt dihydrate | $Na_2.EDTA.2H_2O$ | 7.6 |
| Ammonium sulphate | $(NH_4)_2SO_4$ | 936 |
| Trace element (A & B) | n.a. | 1.25 mL L$^{-1}$ |

The trace elements were divided into two solutions based on its tracing objects. Trace element A solution was made for tracing the anammox elements, whereas trace element B solution for tracing EDTA elements. Therefore, the chemicals for trace element B chemicals (AppliChem GmbH, Germany) were only $Na_2.EDTA.2H_2O$ (8.025 gr) and $FeSO_4.7H_2O$ (9.2 gr). The chemicals (AppliChem GmbH & Carl Roth GmbH + Co. KG, Germany) for trace element A are as stated below:

Table 15: List of chemical substances for trace elements solutions.

| Chemicals | Formula | Weight (g) |
|---|---|---|
| EDTA disodium salt dihydrate | $Na_2EDTA.2H_2O$ | 16 |
| Zinc sulphate heptahydrate | $ZnSO_4.7H_2O$ | 0.43 |
| Manganese(II)chloride heptahydrate | $MnCl.4H_2O$ | 0.24 |
| Calcium sulphate pentahydrate | $CaSO_4.5H_2O$ | 0.99 |
| Molybdic acid sodium salt dihydrate | $NaMoO_4.2H_2O$ | 0.22 |
| Nickel(II)chloride hexahydrate | $NiCl_2.6H_2O$ | 0.19 |
| Sodium selenate | $Na_2SeO_4$ | 0.1 |
| Boric acid | $HB_2O_3$ | 0.0014 |

Both trace elements solutions were made based only on 1 L influent volume. Thus, both trace elements were only added 50 mL for 50 L influent volume. The preparation for both trace elements solutions were like the preparation of synthetic water and both will be kept in 1 L volume bottle. Trace elements solutions need to be stored in the fridge with 0°-4° C temperature, while synthetic water could be kept within room temperature.

### 5.2.4 Analytical methods

Ammonia, nitrite, nitrate, COD, MLSS/MLVSS, and alkalinity were measured using standard methods. Photometrical cuvette tests based on German standards was conducted to find the ammonium, nitrate, nitrite, total nitrogen, TKN, COD and orthophosphate. Mixed liquor suspended solid (MLSS), volatile suspended solid (MLVSS) and $BOD_5$ were checked according to Standard Methods.

All parameters in the experiment were including nitrogen components, pH, temperature, electrical conductivity (Redox) as well as chemical oxygen demand (COD). It will be determined from influent and the effluent from SBR. Nitrogen components was analyzed a measured as ammonium ($NH_4^+$-N), nitrate ($NO_3^-$-N), nitrite ($NO_2^-$-N), COD and total nitrogen (TN) based on German standards using Hach Lange cuvette test kits and Hach Lange spectrophotometer (DR2800, Hach, US). The sample must be filtered using syringe filters with pore size 0.45 $\mu$m (VWR, US). Afterwards, the sample need to be diluted for some nitrogen components and COD. Factor of the dilution was based on the concentration of the related cuvette test kits. TN was consisted of total organic nitrogen (TON) and total inorganic nitrogen (TIN). However, total organic nitrogen could be neglected. While total inorganic nitrogen could be attained from the total of $NH_4^+$-N, $NO_3^-$-N, and $NO_2^-$-N, which was expressed in mgN $L^{-1}$.

### 5.2.5 Calculations

Nitrogen analysis were formed into nitrogen loading rate (NLR), nitrogen removal rate (NRR) and nitrogen removal efficiency (NRE), which based on $NH_4^+$-N and TIN concentration.

$$NLR = \frac{Q_i \times S_{N,i}}{V} \qquad \text{(Eq. 21)}$$

$$NRR = \frac{Q_i \times (S_{N,i} - S_{N,e})}{V} \qquad \text{(Eq. 22)}$$

$$NRE = \frac{(S_{N,i} - S_{N,e})}{S_{N,i}} \times 100 \qquad \text{(Eq. 23)}$$

$Q_i$ is the flowrate of feeding in L day$^{-1}$, while V is the volume of system in one operation day in L. $S_{N,i}$ is the nitrogen related concentration in influent, whereas $S_{N,e}$ is in the nitrogen related concentration in effluent and both were in mgN $L^{-1}$.

Biomass concentration was concluded from mixed liquor suspended solid (MLSS) and volatile suspended solid (MLVSS) measurement based on the Standard Methods 1997

(Clesceri et al., 1998). Biomass from K1 and K3 carriers as well as feed sludge were also occasionally measured and analysed. The measurement for seed sludge usually was done a day after the sludge came from the plant. An additional preparation was conducted started on November 2017, which was drying the filter paper along with the dish in 105° C oven over the night before the MLSS and MLVSS was being carried out.

.

### 5.2.6 Fluorescence in situ hybridization (FISH)

Fluorescence in situ hybridization (FISH) is a very powerful cultivation-independent method for identification of microorganisms in activated sludge and biofilm biocoenosis using oligonucleotide probes, which targeting the ribosomal RNA (rRNA). Quantitative fluorescence in situ hybridization (FISH) was regularly made to identify the types of bacteria followed by light microscopy and digital image analysis. List of probes are given in Table 16. It is also possible to observe the morphology and to quantify numbers of bacteria or the equivalent biovolume using image analysis. The sample for FISH were taken from effluent of SBR and biofilm on K1 and K3 carriers. FISH was conducted once in a month and qFISH was done one time for the activated sludge from Kamen. Anammox, ammonia oxidizing bacteria (AOB) and nitrogen oxidizing bacteria (NOB) were the targeted bacteria's group for FISH and qFISH. Amx820, Amx368, Sca1309 and BS820 are the probes for anammox bacteria. Nso190 is used to targeting AOB, while NIT3 and Ntspa662 for NOB. A general probe EUB338mix is also used to targeting most of bacteria.

*Pre-treatment of sample*

A pretreatment is necessary to aim a homogenous solution, especially with granular sample. After diluting the sample with 1:1 dilution factor, the sample will be put on a vortex body for a general mixing. Then, the sample will be put inside an ultrasonic cleaner (Schallsonic 4000, EMAG AG, Germany) for about 5 minutes to have a cleaner sample thereby the microscopic reading will be effortless.

Table 16: List of used oligonucleotide probes for IFAS-SBR

| Probes | Sequence (5'-3') | Specificity | Dye | Formamide (%) | References |
|--------|------------------|-------------|-----|---------------|------------|
| EUB 338 | GCT GCC TCC CGT AGG AGT | Most bacteria | FITC | 5-55 | (Zarda et al., 1991) |
| Amx368 | CCT TTC GGG CAT TGC GAA | All anammox bacteria | Cy3 | 15 | (Kartal et al., 2006) |
| Amx820 | AAA ACC CCT | Anammox : Candidatus | Cy3 | 40 | (Kartal et al., 2006) |

| | CTA CTT AGT GCC C | Brocadia anammoxidansapos and Candidatus Kuenenia stuttgartiensisapos | | | |
|---|---|---|---|---|---|
| Sca1309 | TGG AGG CGA ATT TCA GCC TCC | Anammox : Candidatus Scalindua | Cy3 | 5 | (Schmid et al.,2007) |
| Nso190 | CGA TCC CCT GCT TTT CTC C | AOB : Protobacterial | Cy3 | 50 | (Mobarry et al.,1997) |
| NIT3 | CCT GTG CTC CAT GCT CCG | NOB : Nitrobacter spp | Cy3 | 40 | (Mobarry et al.,1997) |
| Ntspa662 | GGA ATT CCG CTC TCC TCT | NOB : Nitrospira spp. | Cy3 | 30 | (Daims et al.,2000) |

*Cell fixation and washing procedure*

After the pre-treatment, 2 mL of sample was taken and filled to a microcentrifuge tube (Eppendorf Vertrieb Deutschland GmbH, Germany). Cell of the sample will be separated into liquid (water) and solid (biomass) phase by using a centrifugal device (Andreas Hettich GmbH & Co.KG, Germany) with 4° C and 4,000 rpm for about 15 minutes. It will be lasted about 20-25 minutes notably for sample from effluent of SBR, which were less concentrated. The fixation of cell was conducted by mixing the 500 µL of sample with 1500 µL of 4 % Roti-Histofix. Then, the fixed sample will be incubated ca. 2-24 hours.

After incubation, the fixed cell will be centrifuged with 4°C and 10.000 rpm for about 10 minutes. For a less concentrated sample, the duration would be ca. 15-20 minutes. Then, 500 µL of sample will be washed with 1500 µL of phosphate buffered saline (PBS) 10 % solution (Thermo Fisher Scientific, USA) and centrifuged again. The washing steps were repeated ca. 3 times and for the last step, 750 µL sample will be washed with 250 µL PBS 10 % and 500 µL ethanol 96 % (Carl Roth GmbH + Co.KG, Germany). The washed cell must be stored in freezer and could be used in a few months later.

*Preparation of hybridization and washing buffer*

The buffer for hybridization and washing were prepared correspondingly to formamide concentration of each oligonucleotide gene probe. Formamide is an ionizing solvent which is widely used in molecular biology research for its thermodynamic effects on the DNA double-helix stability. Addition of formamide to aqueous buffers solutions of DNA enables key procedural steps, such as the prehybridization denaturation, the reannealing step and the post-hybridization stringency washes. Thus, the hybridization procedure could be carried out at lower, less harsh temperatures without compromising the overall efficiency and specificity of the hybridization. Formamide solvent isn't used in the washing buffer, but the washing buffer was made subsequently to the formamide concentration on hybridization procedure. Both buffers were prepared in 50 ml tubes using miliQ water. The sodium dodecyl sulfate (SDS) 10 % (Sigma Aldrich, USA) solution must be added on last step to avoid precipitation. For preparation of the washing buffers need to be prepared in a place with less light (see FISH Protocols in appendix D). The composition of hybridization and washing buffer could be seen in Table 17 and 18 below.

Table 17: Hybridization buffer

| Composition | Probe | | | | | | | |
|---|---|---|---|---|---|---|---|---|
| (µL) | Amx368 | Amx820 | Sca1309 | BS820 | Nso190 | NIT3 | Ntspa662 | Chloroflexi |
| NaCl 5M | 360 | 360 | 360 | 360 | 360 | 360 | 360 | 360 |
| Tris/HCl 1M | 40 | 40 | 40 | 40 | 40 | 40 | 40 | 40 |
| Formamide | 300 (15 %) | 800 (40 %) | 100 (5 %) | 400 (20 %) | 1100 (55 %) | 800 (40 %) | 600 (30 %) | 700 (35 %) |
| H₂O | 1300 | 800 | 1500 | 1200 | 500 | 800 | 900 | 1050 |
| SDS 10 % | 2 | 2 | 2 | 2 | 2 | 2 | 2 | 2 |

Table 18: Washing Buffer

| Composition | Probe | | | | | | | |
|---|---|---|---|---|---|---|---|---|
| (µL) | Amx368 | Amx820 | Sca1309 | BS820 | Nso190 | NIT3 | Ntspa662 | Chloroflexi |
| NaCl 5M | 3180 | 460 | 6300 | 2150 | 100 | 460 | 700 | 817 |
| (Formamide %) | (15 %) | (40 %) | (5 %) | (20 %) | (55 %) | (40 %) | (30 %) | (35 %) |
| Tris/HCl 1M | 1,000 | 1,000 | 1,000 | 1,000 | 1,000 | 1,000 | 1,000 | 1,000 |
| EDTA 0,5 M | 500 | 500 | 0 | 500 | 500 | 500 | 500 | 500 |
| SDS 10 % | 50 | 50 | 50 | 50 | 50 | 50 | 50 | 50 |
| H₂O | 45.1 | 47.9 | 42.6 | 46.3 | 48.3 | 47.9 | 47.75 | 47.6 |

All chemicals solution, which were used for the buffers solution above, were prepared by following step:

a) NaCl 5M: 146.1 g of NaCl was mixed in 500 ml miliQ water using magnetic stirrer. Solution was stored in the room temperature.

b) Tris/HCl 1M: 121.14 g Tris (Carl Roth GmbH + Co. KG, Germany) was dissolved in 500 ml miliQ water, adjust pH to 7.0 by adding HCl 5M (Carl Roth GmbH + Co. KG, Germany) and then bring the final volume to 1 liter with miliQ water.

c) EDTA 0.5M: 186.1 g of $Na_2EDTA.2H_2O$ (Carl Roth GmbH + Co. KG, Germany) was dissolved in 1 litre of miliQ water by using magnet stirrer.

d) Ethanol (EtOH) 50 %, 80 % and 96 %: 25 ml, 40 ml, and 48 ml of absolute ethanol consecutively (Sigma Aldrich, USA) were dissolved in to 25 ml, 10 ml and 2 ml of miliQ water respectively based on 50 mL volume tube.

*Application of probe and sample*

Microscope slide (Thermo Fisher Scientific, USA) was prepared before the application by cleaning the slide with ethanol and tissue paper. The slides would be placed on slide rack. Name of probe, the preparation date and sample location would be written on the bottom side of microscope slide. Only 10 µL of washed cell will be added on the well of the microscope slide, afterward the slide would be dried in oven with 46° C. The cell application is preferably to be repeated until three times to have a better result in microscope view. Therefore, it will be better to only added 5 µL of washed cell when the washed cell was too much concentrated.

After drying, the microscopic slide would be dipped into 50 %, 80 % and 96 % ethanol solution orderly for about 5 minutes each solution to dehydrate the cell. Then, the microscope slide would be dried again in oven with 46° C for about 10-15 minutes until the ethanol was totally dried up. Applying the oligonucleotide gene probe need to be done in a place with less light. For a mixed gene probe, 40 µL of corresponding hybridization buffer would be added to each of the aliquoted probe. When EUB338 is one of the mixed gene probes, it needs to be applied firstly on each well of the microscope slide ca. 5 µL volume. Then, the other oligonucleotide gene probe would be added afterwards (details by qFISH Protocols in appendix A).

*Hybridization*

A strip of tissue was placed horizontally in each hybridization buffer tube, and then the corresponding slide was inserted horizontally above the tissue with sample facing upward. The water bath was pre-heated at 46°C, and slides in hybridization buffer were dipped horizontally in the water bath with incubation time varying from 20 to maximum 24 hours. All hybridization steps must be done in the dark.

*Washing procedure*

When hybridization finished, water bath's temperature was increased to 48 °C, the slides were transferred to washing buffer tubes vertically, and incubated in water bath for 15 minutes. After that, the slides were dipped into the distilled water for 5 minutes and dried in the oven at 46°C. When slides were dry, a drop (<5 μl) of anti-bleaching agent Citiflour (Carl Roth GmbH + Co. KG, Germany) was applied on each well. Finally, a glass coverslip was glued on top of each slide. When the glue dried, the slides were ready for microscopy, or can be stored in -20°C for further analysis. All steps must be done in the dark.

### 5.2.7 Digital image analysis

The biomass was visualized by epifluorescence microscopy Axio Imager 2 (Zeiss, Jena, Germany) for quantification and Zeiss LSM 510 confocal scanning light microscopy (CLSM) (Carl Zeiss, Jena, Germany) for better visualization. For quantification, Axio Vision 4.18 software was used for image acquisition, which 20 random fields of views (FOVs) were recorded for each well at 100x magnification and saved in TIFF file format. Afterwards, images were loaded to ImageJ software according to Schneider et al., 2012 (https://imagej.net/Downloads) for the noise reduction, and manually adjusting images threshold for each single image.

Finally, daime 2.1 software (http://dome.csb.univie.ac.at/daime/download-daime) was used for relative abundance measurement due to Almstrand et al., 2014. After image processing, 20 image pairs (TIFF file, size of 8 bits per pixel) was loaded to daime software (Fig. 37). The biovolume fraction function of the software will analyze area of cells in both images (image of specific probe and image of general probe), which expressed by number of pixel. The average bio-volume fraction was calculated based on fraction of area cell detected by specific probe to the entire bacterial cells targeted by general probe EUB338 (Eq. 24). The result includes average relative abundance, pixel sums of each signals (green, red, or blue) and average congruency of the measurement.

$$\text{Relative abundance} = \frac{\sum_{i=1}^{20} A_i^P}{\sum_{i=1}^{20} A_i^B} \qquad \text{(Eq. 24)}$$

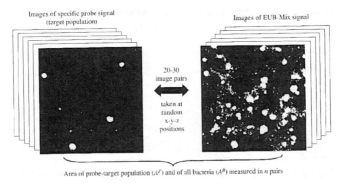

Figure 37: Principle of approach for quantifying the bio-volume fraction of a specific probe (Nielsen et al., 2009)

### 5.2.8 Calculation of weighted average congruency factor

Biovolume analysis function of daime 2.1 software measures average relative abundance of targeted microbial groups and their congruency percentage of images as well. Congruency factor is defined as how signal of the cells detected by a specific probe was congruent with their counterparts, which targeted by EUB338 probe for mix-stained cells in this study. The degree of overlapping between 2 signals (congruency) is expressed in percentage, it means total overlap corresponding to 100 % congruency (Daims et al., 2006). To optimize pre-treatment method for qFISH, the weighted average mean ($\bar{x}$) and standard deviation (S) of congruency were calculated (Eq. 25) for different pre-treatment methods.

$$\bar{X} = \frac{\sum_{i=1}^{n}(x_i \times w_i)}{\sum_{i=1}^{n} w_i} \; ; S= \sqrt{\frac{n \sum_{i=1}^{n} w_i \, (x_i - \bar{x})^2}{(n\text{-}1)\sum_{i=1}^{n} w_i}} \qquad \text{(Eq. 25)}$$

Where $\bar{X}$ is mean, $n$ is number of targeted bacterial groups, $x_i$ is congruency value, $w_i$ is number of images were taken, and S is standard deviation. The highest congruency factor is required to optimize the best pre-treatment method. The congruency factor is also beneficial to check how reliable the estimated relative abundance of targeted group is.

### 5.2.9 16S rRNA high-throughput sequencing

For 16S rRNA high-throughput sequencing, from operaing reactor as well as seeding or biaugmenting sludges were chosen. All samples were stored at $-20°C$ directly after sampling.

To characterize the community composition, genomic DNA was isolated from industrial activated sludge samples using FastDNA SPIN Kit for Soil (MP Biomedicals, Eschwege, Germany) (Dunkel et al., 2016). DNA concentration and purity of the samples were measured photometrically using the Nanodrop 2000 Spectrophotometer (Thermo Scientific, Dreieich, Germany). The specific primer pair of 515F (5'GTGCCAGCMGCCGCGGTAA-3') and 806R (5'-GGACTACHVGGGTWTCTAAT-3') with the barcode will be used to amplify the V4 regions of total DNA. Sequencing libraries will be generated using TruSeq® DNA PCR-Free Sample Preparation Kit (Illumina, USA) and index codes will be added. The library quality will be assessed on the Qubit@ 2.0 Fluorometer (Thermo Scientific) and Agilent Bioanalyzer 2100 system. At last, the library will be sequenced on an Illumina HiSeq2500 platform and 250 bp paired-end reads will be generated. Sequences analysis will be carried out using QIIME. The raw reads will be sorted based on barcodes into different samples, then the adapters, barcodes, and primers will be trimmed off. The paired-end reads will be merged using FLASH V1.2.7, the sequences shorter than 170 bps will be discarded. Chimera checking will be performed by USEARCH, then all sequences will be clustered into one fasta file for OTU picking with the $\geq 97$ % similarity and taxonomy assignment against SILVA 128 released reference database.

### 5.2.10 Batch evaluation during temperature reduction phase

The in-situ batch evaluation was conducted inside the same SBR along with all of the operating conditions, especially pH level was kept on ± 7. The operating cycle was started on 6 am and finished on 6 pm (12 hours). Sample from influent tank was counted as the initial time sample and sample from discharging time as the final sample. The second sample was taken on the $2^{nd}$ hour (08:30-09:00) because when the 1st hour began, the aeration and mixing phase was just also started. Only 30 mL volume of sample from SBR was taken as to avoid a sharp decline of water level in SBR and directly stored in the freezer. The day after, the sample was thawed out from the freezer to be used for nitrogen components (including COD) and biomass concentration measurement. Thus, the sample was divided into half. The batch test was especially carried out along with the declining temperature. First batch test was carried out at 25° C, then 20° C; 15° C; 10° C; 5° C. Two samples each from K1 and K3 carriers were also taken out once during the batch test for dry matter measurement.

## 5.3 Results and Discussions

### 5.3.1 Reactor programming

IKS program, Aquasoft, was acted as the controller, indicator and storing the data. The parameters, which were stored and displayed by Aquasoft, were including date, time, redox (mV), temperature (°C), pH (Acid-Base) and dissolved oxygen concentration (mg L$^{-1}$).

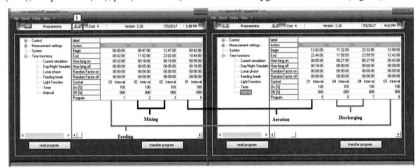

Figure 38: Interface preview of Aquasoft

Number of intervals was based on the position of each related power outlets, as it can be seen in Fig. 38. Alarm for pH was maintained by option 'Control' and the position of sensors by 'Measurement setting'. Before programming, the display on IKS needs to be changed to option "Regelung". Then by clicking the right arrow (1), Programming option will be shown. First step of programming the Aquasoft starts by clicking the 'read program' box. Afterward, the desired value for each interval was given and checked thoroughly one by one. At last, the realization of the given program to Aquasoft was finished by clicking 'transfer program' and it will be also shown on IKS display. After the transfer was done, the date and time of the SBR system need to be also given by choosing option "System". The program will be running after choosing option "Kontrolle".

Duration for recording the data could also be modified by option "Timer" and was lasted mostly for 15 minutes recording. However, the storage by Aquasoft was not high. Thus, the data reading and saving in excel need to be done at least once every two days. On 25$^{th}$ of September 2017 (Day 103), the recording duration was made into 10 seconds to see the fluctuation of dissolved oxygen concentration. Data acquisition and saving starts by clicking the right arrow (1) to change it into read data. After data reading was finished, the data could be transferred and saved as excel file for an easier data's execution. Programing and reading the data was done on portable computer.

### 5.3.2 Reactor maintenance and performance

SBRs were initially set up and run with HRT at 1.5 to 1.7 days. The experiments in the phase I (lasted for 70 days for each reactor) were developed with varied temperature from 30°C to 24°C and reduced influent ammonium nitrogen concentration of 180 mgN $L^{-1}$ to 90 mgN $L^{-1}$. Three crucial parameters, which need to be always kept in the desired value, were pH, DO concentration and temperature. The accommodating pH value for anammox bacteria, is in the range 6-8. The real-time dissolved oxygen (DO) concentration during was controlled below 1 mg $L^{-1}$ using online sensor. The aeration was operated in 26 cycles with varied duration between 3-5 minutes and 15-20 min for mixing for about 80 days for both reactors. During start up periods of both plants, different aeration patterns have been tested by changing the times and intensity with regulating a micro air membrane pump. The reactor 1 reached to a stable removal efficiency of more than 50 % after 25 days while reactor 2 failed to reach this threshold (Fig. 39 and 40). By further integration of suspended biomass into moving bed media in IFAS, the nitrogen removal efficiency reached to 91.24 % in the 48th working day of the reactor 1 which is 2-fold higher than the maximum nitrogen removal rate in reactor 2 with combination of granules and activated sludge. Besides, biomass washout of floating granules was experienced during the beginning of the operation of granular sludge reactor. The highest nitrogen removal rate (NLR) for IFAS reactor as observed at the day 26 with 0.1 kgN. $m^{-3}$ $d^{-1}$ while for granular sludge reactor was observed at the day 60 with 0.08 kgN. $m^{-3}$ $d^{-1}$. Since in the phase I, IFAS-SBR (reactor 1) showed a better performance, it was selected for further investigations to find the effect of cold temperatures. This the reactor temperature in second phase were reduced to 5°C.

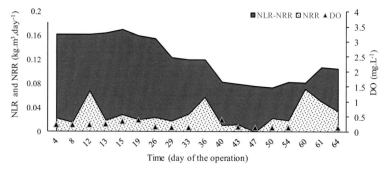

Figure 39: Nitrogen removal efficiency, nitrogen loading rate (NLR) and nitrogen removal rate (NRR) for the reactor 2 mixing granular and activated sludge

Nevertheless, the operation of IFAS-SBR was divided into three phases, with a transition in between Phase I and II, due to a sudden unexpected pH drop (Fig. 41). The temperature was started from 30°C and controlled for Phase I and II reducing to 21°C. Ideal temperature condition for growing the anammox bacteria was taught to be in the range 26°–28° C (Thamdrup & Dalsgaard, 2002). The Phase III was started after reaching the temperature to 21° C, which was on day 141 and ended on day 187 with 5° C (Fig. 42).

The beginning of SBR's operation until day 75 is Phase I, then the transition was begun due to the problem in SBR and lasted until day 91. On the first operation week for IFAS-SBR, the highest nitrogen removal efficiency (NRE) based on ammonium concentration (Fig. 40. (a)) was only until 68 %. The highest NRE in whole phases was 99.92 % on day 28. The highest nitrogen removal rate (NRR) were 0.14 kgN m$^{-3}$ day$^{-1}$ based on NH$_4^+$-N concentration with NRE 96.6 % and 0.15 kgN m$^{-3}$ day$^{-1}$ g.m$^3$/day with NRE 83.6 % based on TIN concentration. However, the declination of NRE in Phase I happened twice, which was on operational day 37 and 58. Phase II was lasted day 135 and acted as a recovery phase. The highest NRE after the transition in Phase II was only 78.15 % on the day 91. However, the NRE was dropping several times. Once was happened on day 121 and assumed due to the decline of temperature to 25°C. Then, it was occurred again on day 131 and assumed due to the cycle's change, which was changed into 3 cycles with 8 hours for each cycle duration on day 128. In Phase III, the highest NRE for ammonium-nitrogen was 63 % with temperature 18°C. At the end of operation day, NRE could only reached 11.7 % with 5°C temperature. Total inorganic nitrogen is the total of NH$_4^+$-N, NO$_3^-$-N and NO$_2^-$-N.

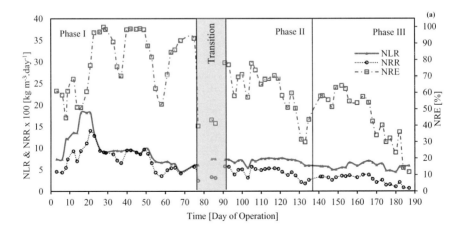

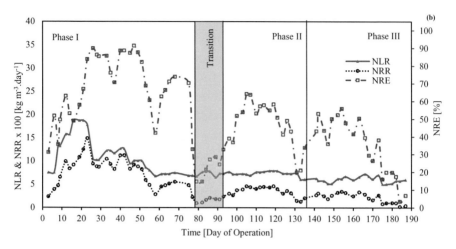

Figure 40: Nitrogen removal (a) based on ammonium (b) based on total inorganic nitrogen.

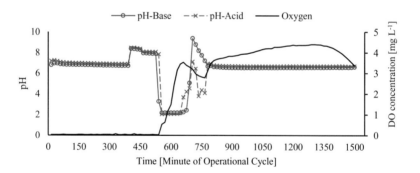

Figure 41: pH drops on 28th of August 2017

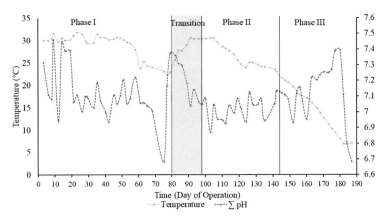

Figure 42: The average pH and temperature

### 5.3.3 Temporal variations of inorganic nitrogen concentrations

The $NH_4^+$-N concentration in influent and effluent was regularly monitored and controlled. However, the $NO_3^-$-N and $NO_2^-$-N concentration sometimes are higher than it should be as seen in Fig. 43 (a). In early operation day (before day 42), except when there was a problem, the $NO_2^-$-N concentration was until 44.9 mg $L^{-1}$. It was assumed that there was a partial nitritation or complete nitritation in influent because the influent was too long settled and was not mixed. Therefore, the influent was changed once per week and was mixed on afternoon for every working work day. Still, the highest $NO_2^-$-N concentration in influent was not over the limit of $NO_2^-$-N concentration for anammox process.

Reduction of $NH_4^+$-N concentration in influent was made during Phase I to see how it will influence the nitrogen removal efficiency. It was reduced 10 % from each previous $NH_4^+$-N concentration in synthetic water preparation. As seen in Fig. 43 (b), there was an increase of $NO_2^-$-N concentration in effluent on day 91 in transition phase, while $NH_4^+$-N concentration was lower. Due to the attempt of a good nitrogen removal efficiency while the left over anammox bacteria was revived, the aeration time was made to 10 minutes on day 89. Thus, $NO_2^-$-N concentration was reached till 42.9 mg $L^{-1}$ in effluent.

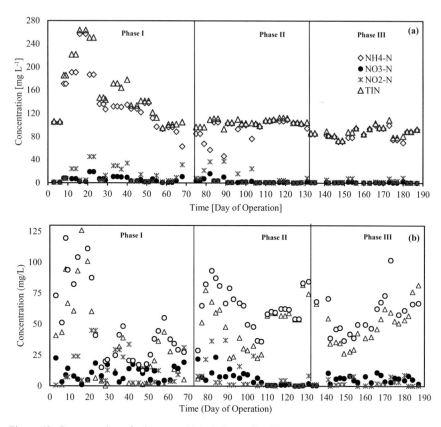

Figure 43: Concentration of substances (a) in influent (b) effluent.

### 5.3.4 Intermittent aeration regimes

Various duration of aeration was attempted in this experiment. DO concentration needs to be very low ($<1$ mg $L^{-1}$) as it will cause inhibition to anammox process. However, partial nitritation needs some amount of DO concentration. Therefore, an aeration pattern was applied to the IFAS-SBR to aim a more efficient aeration for partial nitritation-anmmox process. For IFAS-SBR after the day 25 (start-up period) the averaged DO (per day) for nitrogen removal was determined to be between $0.2 - 0.7$ mg $L^{-1}$ (Fig. 44) and maximum real-time DO at the end of aeration cycle was set 1 to 1.2 mg $L^{-1}$.

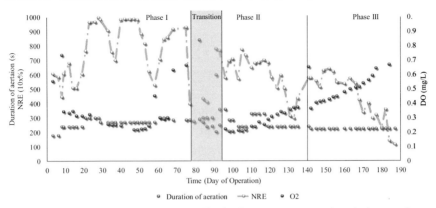

Figure 44: Aeration pattern and average dissolved oxygen concentration during aeration phases.

The first aeration phase was begun after the feeding. Then, the mixing phase started and the aeration-mixing phase (react phase) will be repeated to 26 cycles. The total time for one phase of aeration-mixing was between 23-24 minutes, which aeration only took ca. 20 % of the total. The longest aeration in Phase I was 300 seconds and on day 28 the DO concentration was until 1.22 mg L$^{-1}$. However, the NRE based on ammonium was even 99.92 % as shown in Fig. 40 (a) and for NRE based on total inorganic nitrogen was 86.11 %. Therefore, a better nitrogen removal efficiency could be reached even with DO concentration >1 mg L$^{-1}$. In the transition phase, the aeration lasted even until 10 minutes on day 91 to increase the nitrogen removal rate and efficiency. Nevertheless, the DO concentration at the time of highest system performance was only ca. 0.2 mg L$^{-1}$ as shown in Fig. 44.

### 5.3.5   Influence of temperature

Reduction of the temperature was tried for several times, both in Phase I and III. In Phase I, it was reduced with a regular water bath. While in Phase III, the reduction of temperature was controlled by a thermostat. The effect of temperature on the nitrogen removal could be seen better on batch test. As stated before, batch test was conducted 5 times with different declined temperatures. There were declinations of NH$_4^+$-N amount in IFAS-SBR in between 2$^{nd}$ and 3$^{rd}$ hours of cycle (see Fig. 45 (a)). It occurred in almost every degree of batch test, except at 5°C. It might be that the partial nitritation and anammox activity were taking place properly in the early cycle hours due to fresh influent. As shown in Fig. 45 (b) & (c)., the amount of NO$_2^-$-N had declinations also at 2$^{nd}$ and 3$^{rd}$ hours of cycle, while NO$_3^-$-N had inclinations. It could be assumed that $NH_4^+$-N was oxidized partially into $NO_2^-$-N and both $NH_4^+$-N & $NO_2^-$-N was converted to $NO_3^-$-N & $N_2$ by anammox bacteria. This could also be seen on the COD amounts at $2^{nd}$ and $3^{rd}$ hours of cycle (see Fig. 45d), which was increased due to the consumption of oxygen in partial nitritation process.

As for the rest hours of cycle for $NH_4^+$-N was fluctuating but had tendencies to decline towards the end of the cycle. However, it was increased at 5°C and 25°C. $NO_3^-$-N, $NO_2^-$-N and COD had inclinations at 5°C. While at 25°C, they had declinations. It was considered that at 5°C, the partial nitritation-anammox process in IFAS-SBR at end of the cycle (settling time) could not be efficiently controlled. Also, the adaptation of anammox bacteria at 5°C need longer time for better biomass acclimation and adaptation, which might be at least a week before a batch test was conducted. Whereas at 25°C, the inclination of $NH_4^+$-N at the end of cycle was occurred at discharging time ($12^{th}$ hours). Even with the declination of other substances ($NO_3^-$-N, $NO_2^-$-N & COD). It was suspected that denitrification could play a key role in the system. An unusual inclination of $NO_3^-$-N happened at 15°C on $6^{th}$ and $8^{th}$ hours of cycle, ca. 100-140 mg $L^{-1}$. At the same hours, the COD had declination and amount of $NO_2^-$-N was also low. It could be concluded that at those hours anammox process was dominant. However, several duplicate samples at 15°C should be taken as to know the real reason. The behavior on each degree could also be influenced by the reduction of initial $(NH_4)_2SO_4$ concentration in synthetic water. As shown in Fig. 45 (a), $NH_4^+$-N was reduced at 25°C through 20°C. However, $(NH_4)_2SO_4$ concentration in synthetic water was increased again to support the growth of anammox bacteria from 15°C to 5°C. Despite the increase of nutrient, the effect of low temperature was higher on system performance. Hence, the drop of nitrogen removal efficiencies happened in Phase III. Nevertheless, the most effective temperature was at 20°C based on the substances ($NH_4^+$-N, $NO_3^-$-N, $NO_2^-$-N and COD) concentrations.

### 5.3.6 Biomass concentration

The existence of the bacteria (biomass) could be measured through dry matter test, mixed liquor suspended solid (MLSS) and mixed liquor volatile suspended solid (MLVSS). MLSS represents the total concentration of suspended solid in activated sludge, including organic or inorganic suspended solids. Whereas, MLVSS indicates the organic bacterial cells. The standard value of MLSS for leachate sludge are ± 7 g $L^{-1}$ (Tsilogeorgis et al., 2008).

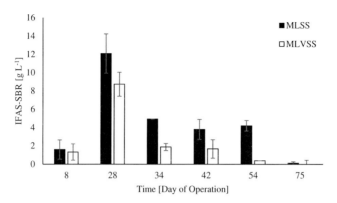

Figure 45: Dried matter test (MLSS and MLVSS) of activated sludge (flocs) in mixed liquor from Phase I from IFAS-SBR

As shown in Fig. 46, the feed sludge for Phase I were from ZDE and ZDA (Germany). The initial feed sludge was from ZDE (Gelsenkirchen, Germany) and the sludge in IFAS-SBR was mixed from sludge ZDA (Aachen, Germany) since the day 24 (08th of July 2017). On day 28, MLSS and MLVSS values were at the highest ca. 12.08 ± 2.15 g $L^{-1}$ and 8.73 ± 1.32 g $L^{-1}$ respectively (see Fig. 46).

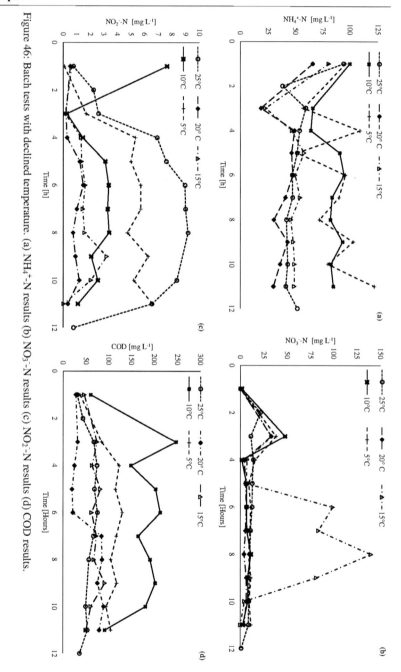

Figure 46: Batch tests with declined temperature. (a) NH$_4^+$-N results (b) NO$_3^-$-N results (c) NO$_2^-$-N results (d) COD results.

Figure 47: Carriers K1 and K3 from IFAS-SBR on day 187.

Carriers could be used to avoid the inhibition of anammox activity due to the residual of DO after partial nitritation process. In this experiment, two types of carriers were used. Kaldnes 1 (K1) carriers were received from municipal waste water in Hattingen (Germany), while Kaldnes 3 (K3) were new carriers. On the day 187, the total of K1 carriers were 471 and K3 were 130 (Fig. 47). In term of the biomass concentration, which were in solid form, suspended solid (SS) and volatile suspended solid (VSS) tests were also conducted for both types of carriers.

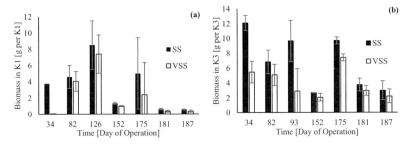

Figure 48: SS and VSS results from biomass concentration on (a) K1 carriers (b) K3 carriers.

In Fig. 48 (a), the highest value of SS and VSS in K1 were on day 126, which were $8.5 \pm 2.99$ gr SS per K1 carrier and $7.4 \pm 2.35$ gr VSS per K1 carrier. While for K3, the peak of SS was on day 34 for about $12.09 \pm 1$ gr per K3. However, it was different for VSS value, which was on day 175 ($7.41 \pm 0.47$ gr per K3 biofilm carrier). It could be assumed that on day 34, there were lot of inorganic suspended solids due to the mixed sludge from ZDE and ZDA (Germany). Whereas on day 175, a new seed sludge from Kamen (Germany) was added for about a week before, which was on day 166. MLSS and MLVSS for effluent were also measured. Nevertheless, the results were mostly negative, which indicates that there was no biomass in the effluent from IFAS-SBR. One source of the difference between SS and VSS is also due to the boundEPS (non-soluble EPS) content in the biofilm solid matrix fixed on the carriers. This has been shown to contribute around 10 % of total solid content as it was discussed in the last chapter (Azari et al., 2018).

### 5.3.7   FISH and qFISH

FISH and qFISH were used to identify the types of bacteria. FISH was done several times with different gene probes for the sludge from the reactor (Fig. 49-51).

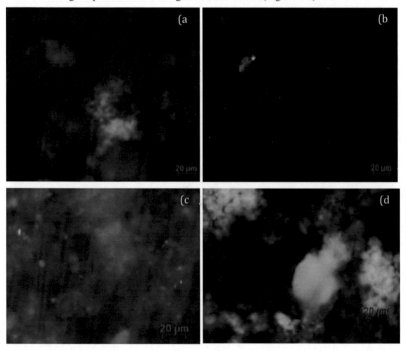

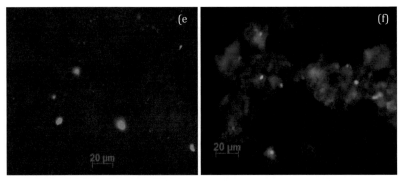

Figure 49: FISH image results from (a) IFAS-SBR on day 51 with BS820 target probe (b) IFAS-SBR on day 51 with Chloroflexi target probe (c) IFAS-SBR on day 51 with NIT3 target probe (d) K1 carriers on day 23 with Sca1309 target probe (e) K1 carriers on day 54 with Amx820 target probe (f) IFAS-SBR on day 187 with Amx368 target probe.

The targeted FISH probes were anammox bacteria, AOB and NOB, which mixed with general probe EUB338. As it was described before, the procedures of FISH and qFISH are similar, which need to be done in 2-4 consecutive days. In microscopic view, the green signal indicates the general bacteria group (EUB338), while the red signal shows the targeted specific probes. It can be seen in Fig. 49 that for the specific probe type like BS820, Sca1309 (anammox bacteria group). Therefore, it was assumed that there were very few or might be no populations for the specific types of anammox bacteria *Candidatus* Scalindua (Sca1309), *Candidatus* Scalindua wagneri and *Candidatus* Scalindua sorokinii (BS820) due to infrequent existence of *Candidatus* Scalindua spp. in wastewater since its domain is non-polluted environment. Similar case was also occurred for NOB probes, NIT3. Therefore, in the later FISH tests only Ntspa662 probe was used for identifying *Nitrospira* (NOB) but only for a few samples, a very low abundance of NOB has been detected using this probe.

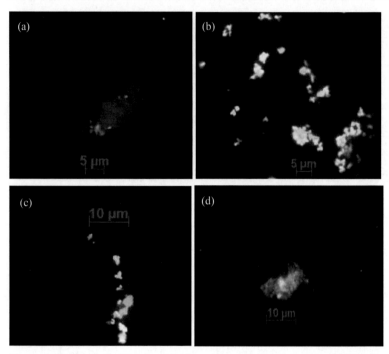

Figure 50: FISH results from day 175 after adding granular sludge from WWTP Kamen used for bioaugmentation at the last phase during reduction of temperature (a) sample from activated sludge flocs of IFAS-SBR with Ntspa662 target probe (NOB probe) (b) sample from K3 biofilm and (c) sample from K1 biofilm with Nso190 target probe (AOB probe) (d) and with Amx382 target probe (anammox bacteria probe).

The results from qFISH were given as biovolume abundance with respect to the volume fraction of specific probe, which was obtained by series of images with congruency >95 %. Analysis of relative abundance of targeted microbial groups was done from at least 20 random FOVs. The specific target gene probe for qFISH were only Amx368, Nso190 & Ntspa662, which were also mixed with the general probe EUB338. Previously, the Amx820 probe was always used to detect other type of anammox bacteria beside Sca1309 probe due to its frequent presence in anammox unit process. The Amx820 probe is specifically for *Candidatus Brocadia anammoxidansapos* and *Candidatus Kuenenia stuttgartiensisapos* identification. However, those two types of *Candidatus* were not able to be detected after the problem was aroused (Phase II). Therefore, Amx368 probe was selected for qFISH and later

FISH test since it covered all the type of anammox bacteria. Ntspa662 probe was also used considering that *Nitrospira spp* was not able to be detected by NIT3 probe. The qFISH itself was conducted in Phase III due to the availability of time and IFAS-SBR condition.

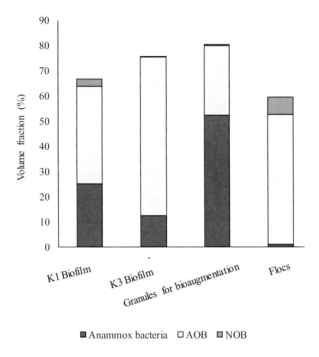

Figure 51: Estimated relative abundance from qFISH on day 166 for floccular sludge and carrier biofilms samples taken directly from the reactor and granular sludge of bioaugmentation taken fresh from the full-scale plant Kamen for before adding to the reactor

As shown in Fig. 51, the most presence bacteria were AOB for all the samples except sample from Kamen. It's to be expected as the sludge from Kamen was collected in form of granular sludge, thus the dominance from anammox bacteria. For K1 and K3 carriers, anammox bacteria was placed second place. While in sample from IFAS-SBR, anammox bacteria was only 1,7 %. It could be assumed that the anammox bacteria could only live in the biofilm carrier and not in the flocs form, since the sample from IFAS-SBR were the flocs form and anammox bacteria was presence in at least 25.1 % volume fraction in K1 & 12.6 % volume fraction in K3 carriers. Although the size of K3 was bigger than K1, the volume fraction for K3 was slightly less than in K1. It could be considered that in K3 only a thin layer of biofilm was retained since the carriers were mobile and it could have resulted in biomass loss.

As it was disclosed before, AOB will outcompete NOB when the temperature is higher than 20°C and NOB will be dominated when the temperature is dropped lower than 15°C. However, it wasn't the case for IFAS-SBR since the AOB was still abundant in the lower temperature (10°C) when this qFISH was conducted on day 175. The majority of AOB in the seed sludge, Kamen's sludge, could be considered as one of the reasons since the DO concentration ($\pm 0.06$ mg $L^{-1}$) did not also gave an impact to it. This circumstance could also be occurred because the after effect of the pH-drop, hence the unsuitable condition in the reactor for anammox bacteria to grow in the floc formation. On account that the anammox bacteria is extremely difficult to cultivate even in the pure culture could as well be one of the causes. Moreover, the temperature was very low on day 175 and it was not a suitable temperature condition also for anammox bacteria to grow or to sustain.

The AOB probe was Nso190, NOB was Ntspa662 probe and AnAOB was Amx382 probe. In Fig. 50, it could be seen the abundance of AOB rather than the other two bacteria groups for the sample from reactor, which were in form of biofilm carriers K1 and K3 and activated sludge. However, anammox bacteria was the dominant group for Kamen's sludge. This could be roughly assumed based on the form of Kamen's sludge itself, which was in granular form (see Fig. 52) since anammox bacteria was mostly found in granular sludge (Sekiguchi et al., 1998).

Fig: 52. Granular sludge from WWTP Kamen (left) and ZDA, Aachen, Germany (right).

### 5.3.8 High-throughput Illumina sequencing of 16S rRNA genes

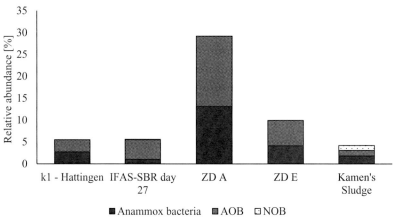

■ Anammox bacteria  ■ AOB  □ NOB

Figure 53: 16s rRNA sequencing from day activated sludge of the reactor at the day 27 as well as sludges used for seeding and bioaugmentation.

In Phase I, a 16s rRNA sequencing was conducted at other as shown in Fig. 53. The results from the sequencing shown that AOB was already a major bacteria group in the IFAS-SBR, even on the seed sludge from ZDA and ZDE (Germany). Whereas by biofilm carriers from Hattingen, the AOB and anammox bacteria were in evenly balanced. However, each of them was less than 3 % and the the relative abundance of anammox from other landfill leachate treatment plants were only 17.2 % in total. It could be presumed that the AOB also outcompete the anammox bacteria since it was already dominant in the three seeds sludge. Moreover, the doubling days for anammox bacteria was 11 days (Jetten et al., 2001). Also, AOB has a higher growth rate when the temperature was higher than 20°C, which is also an ideal growth condition for anammox bacteria. These facts were also applied on the sample sludge from Kamen, which was collected from the plant directly and the sludge itself was not used for bioaugmentation until day 166. All bacteria groups were in similar relative abundancy on the day 27, then NOB was completely outcompeted by AOB and anammox bacteria in phase III (Fig. 50).

### 5.4 Conclusion

The experiment was intended to be a partial nitritation-anammox based process, although in the fact there might be denitrification process inside, hence simultaneous partial nitritation anammox denitrification (SNAD) process could be the existent process for this IFAS-SBR. The intermittent aeration pattern itself does not give a direct impact towards the inhibition of anammox growth. However, it gave better results for nitrogen removal efficiency and the condition's balance in the reactor between the oxic and anoxic section. The most optimum intermittent aeration pattern was found for an aeration duration of 5 minutes and dissolved oxygen concentration during oxic phase to be controlled at 1.2 mg $L^{-1}$. It was also helped the nitrogen removal efficiency (NRE) to revive after the pH-drop. The optimum pH values were ~7 when the NRE was >95 %. The highest NRE itself happened in phase I on day 28 after granular sludge from ZDA was bioaugmented into the reactor, which means sludge from ZDA has exceptional impact.

The direct effect on the system was actually came from temperature. When the temperature was reduced, it could be assumed that anammox bacteria will need longer adaptation time towards lower temperature. On that account, it's preferably to reduce the temperature one degree every 3 days or even a week. It will also be very recommended to not reduce the nutrient content when the temperature was reduced to extreme degrees (<15°C), thereby the anammox bacteria could receive enough nutrient to sustain. The optimum temperature was 28°-30°C when the NRE was >95 % and below 20°C, the system will need longer time to adapt. During the transition phase, the reactor was left to have a self-recovery for about less than 3 weeks and afterwards new seeding activated sludge from leachate treatment plant ZDE in Gelsenkirchen (Germany) was added to regain the system performance. However, a full self-recovery was not able to be achieved and NRE could only reached less than 78 % even after the new sludge was added. If this similar problem (pH-drop) happened in the future, a thorough cleaning of the reactor and then fresh start-up operation is preferably. Otherwise, additional of new good seed activated sludge needs to be done as soon as possible. Through the transition and recovery phases, the biofilm carriers were able to be the media, where some AOB and anammox bacteria and very few NOB could be still attached. Therefore, using biofilm carriers are really good in maintaining the presence of microbes as well as the growth of anammox bacteria since it provides anoxic space. At the same time, it was also supporting the presence and growth of AOB simultaneously. Nevertheless, AOB were majorly found in floccular sludge compared to biofilm and NOB were successfully suppressed in both activated sludge and biofilm samples during initial phase. Therefore, it is highly recommended to apply the support of plastic carriers for PN-A and SNAD process subsequently.

# 6    Conclusion and outlook

## 6.1    Lessons from data analysis of a full-scale application

The performance of a biological treatment plant treating high-ammonium-strength landfill leachate in 2006–2015 was studied. The plant was upgraded combining activated sludge with activated carbon biofilm. The effect of C:N ratio on consistent and stable anammox was discussed, the long-term risk assessment was conducted, and the bacteria community was analyzed. The main lessons for further application and optimization of anammox-based applications are:

i.    Anammox are better enriched by combining activated sludge and activated carbon biofilm. For this, granulation is one of the alternative biofilm-based technologies to increase solid retention time in the reactors. With this process as final polishing step after conventional activated sludge process, a high nitrogen removal of 94 % and long-term stable operation can be reached. Besides, the energy efficiency, methanol consumption and excess sludge was significantly reduced using granulation as final polishing. This idea was tested and regulated successfully in a leachate treatment plant.

ii.    Marine anammox bacterium *Ca.* Scalindua was found remarkably in this plant. Further inoculation and enrichment of this strain can be useful due to adaptability to lower temperature and higher salinity.

iii.    A risk assessment tool based on the long-term actual data was developed and introduces. For further study of similar anammox-based full-scale plants in order to choose the best technology using this comparison tool for long-term study is beneficial.

## 6.2    Lessons from the modeling approaches

First, a biofilm model based on ASM1 was proposed for denitrification and anammox in anoxic conditions and the model was calibrated and verified with short-term batch experiments for different feeding conditions. The model could mimic the single anammox process, denitrification and simultaneous anammox and denitrification (SAD) well. In this model novel stoichiometric matrix was initiated for anammox process and two microbial subgroups were considered for anammox. The idea of differentiating between major anammox species in the model was used to estimate unknown kinetics and physiological parameters of two common genera of anammox planctomycetes. This was only possible after a full sensitivity analysis and parameter identifiability analysis of model parameters and recognizing the identifiable parameters using collinearity index calculations. The maximum specific bacterial growth rates ($\mu_{max}$) for *Candidatus* Brocadia anammoxidans and *Ca.* Scalindua sp. were estimated at 0.0025 h$^{-1}$ and 0.0048 h$^{-1}$ respectively. Decay rate of *Ca.* Brocadia anammoxidans was estimated at 0.0003 h$^{-1}$ which is 15 % higher than decay rate of species belonging to *Ca.* Scalindua.

Afterwards two comprehensive models including and not including the kinetics of SMP and EPS in a granular activated carbon-assisted biofilm reactor were developed. Two models were built-up to explain not only the fate of the substrates and nitrogen transformation in the biofilm reactor but also to express the relative abundances of independent microbial groups. Nevertheless, the model was improved after the addition of EPS and SMP kinetics and the real-time production of EPS and SMP can be predicted as well. This improvement in terms of prediction of nitrogen components was not significant compared to improvement of prediction of microbial relative abundances. Anammox bacteria were the most dominant species with 50 to 60 % percentage in biocoenosis of the biofilm and this was predicted with a good accuracy with the model with EPS better than the model without EPS kinetics.

The main lessons from long-term modeling approach were:

i.   There is a long-term balance and stability between the different microbial and non-microbial groups involved which is highly important in system stability and performance. No specific correlation is hypothesized to link between the system performance in terms of efficiency and stability with the proportion of active and inactive microbial groups. However, the results proved a significant correlation between the nitrogen surface load with the prediction of SMP products and the particulate matters in the biofilm especially the EPS, heterotrophic denitrifiers and inert concentration.

ii.  The main question formulated in the objectives of this study was to answer whether and when it is necessary to define and apply such a complicated biofilm model using the EPS matrix definition. Comparison of results of two models regarding the system performances and nitrogen degradation demonstrated that there is no privilege of the model with EPS to the model without EPS in terms of design and operation proposes.

However, the model without EPS slightly underestimate the heterotrophic denitrifiers fraction due to neglecting the growth kinetic of heterotrophs on SMP, and it suggested a lower accuracy for predicting values and trends of relative abundances of anammox bacteria, AOB and NOB.

iii. Considering the rapid development of biofilm-based technologies, the outlook of developed biofilm mathematical model is specializing and providing a useful tool for researchers to investigate and better understand the microbial consortia and population dynamics in a biofilm reactor compared to the traditional activated sludge models. This modeling approach would also bring benefit for the plant operators by creating a better insight into the real-time microbial composition within the full-scale reactors. Besides, the application of a calibrated and validated mathematical model to examine microbial consortia dynamics requires less technical and laboratory effort and fiscal investment compared to applying sophisticated molecular techniques.

iv. Many studies proved the mutual co-existence and collaboration between denitrifying bacteria and anammox planctomycetes in biofilm systems. Some of these studies quantified the relative and/or absolute abundance of these two groups of bacteria using variety of molecular technique or using numerical simulation. A brief comparison of various molecular techniques to give an idea about relative abundance of microbial groups is given in Table 19. Results from Table 19 indicated that the proportion of heterotrophs in the nitrifying biofilm was higher than in the anammox biofilm. The first reason might be that the growth rate of autotrophic nitrifying bacteria is higher than that of anammox bacteria this fact is supporting the growth of heterotrophs (Makinia, 2010; Zhang et al., 2017). Second, the yield coefficient of SMP production on the growth of nitrifiers ($f_{UAP}^{AOB} = 0.17$ gCOD g$^{-1}$N and $f_{UAP}^{NOB} = 0.16$ gCOD g$^{-1}$N) is higher than similar values for the anammox ($f_{UAPA}$,max$=0.05$ gCOD g$^{-1}$N). Third, the range identified for the decay rate of nitrifiers ($0.002$ to $0.2$ d$^{-1}$ for NOB and $0.015$ to $0.15$ d$^{-1}$ for AOB) is faster than the anammox decay rate ($0.0011$ to $0.0081$ d$^{-1}$), therefore they produce more BAP than amammox bacteria in a biofilm system. So, the quantity of SMP products will be higher for nitrifying bacteria rather than anammox process, which acts as alternative organic carbon source for the growth of heterotrophic denitrifiers.

Table 19: Comparison the active biomass fraction of anammox and heterotrophs

| Type of biomass | AOB ( % ) | NOB ( % ) | Anammox ( % ) | Heterotrophs ( % ) | Quantification method | Reference |
|---|---|---|---|---|---|---|
| Granular sludge covered on GAC, full-scale leachate treatment plant (oxygen limited conditions) | 3-5 | 2-3 | 50-60 | 20-25 | qFISH and modeling | This study |
| Anammox biofilm in lab-scale reactor | na | na | 77 | 23 | qFISH | (Ni et al., 2012) |
| Anammox granular sludge | na | na | 60 | 10 | Model based study | (Liu et al., 2016) |
| Anammox biofilm in a full-scale plant | na | na | 64 | 33 | q-PCR | (Ke et al., 2015) |
| Autotrophic nitrifying biofilm in lab-scale for PN/A process | 22 | 28 | na | 50 | qFISH | (Kindaichi et al., 2004) |
| Nitrifying biofilm in lab-scale | 23 | 50 | na | 27 | qFISH | (Nogueira et al., 2005) |
| Nitritation-anammox biofilm on immobilized plastic carrier in lab-scale reactor | <20 | na | 20 - 65 | na | qFISH | (Almstrand et al., 2014) |
| Nitritation-anammox granular sludge in lab-scale reactor | 22 | na | 29 | na | High throughput DNA sequencing | (Chu et al., 2015) |
| Anammox sludge in a pilot-scale plant | 2.7-3 | 0 | 4.5 | 86.4-87.3 | Direct 16S rRNA gene sequencing | (Rosselli et al., 2016) |

As recommendations for further model improvement, it is important for future works to measure the missing initial fractions which are sensitive parameters for the prediction of relative abundances of bacteria and to monitor the data of the dissolved oxygen regularly which is highly important for the model validation. Assuming the real-time variations in the inflow, characterization of the influence of changes in the solid retention time, temperature and pH into the model kinetics would be also of great importance for further works. Besides the adsorption kinetics for simultaneous GAC process in the reactor can be further improved to describe more realistically the mechanisms of the production and degradation of carbon and nitrogen components. It means that more deterministic approaches should be integrated with biokinetic processes.

## 6.3 Experimental approaches: how can we go towards hybrid applications?

### 6.3.1 Optimizing the aeration regimes

To have a greater nitrogen removal efficiency (NRE) with low cost and energy, the anammox process is integrated with other processes, such as partial nitritation or simultaneous nitrification and denitrification. It was proven that the integrated processes could be conducted in sequencing batch reactor (SBR) with integrated fixed-film activated-sludge (IFAS) hybrid process. IFAS is known also as hybrid biological reactor, which combines the features of suspended flocs and attached biofilm growth processes by incorporating the specially designed biomass carriers (biocarriers). This addition of carriers will increase the biomass inventory subsequently also the treatment capacity of reactor due to the attached and populated biomass on the carriers (Singh & Kazmi, 2016). The biocarriers could also provide a longer SRT to facilitate slow growing nitrifying growth and promote nitrification (Kim et al., 2010).

The effect of intermittently regulated aeration pattern and temperature reduction on system performance were studied and microbial community structure were analysed. A single lab-scale IFAS-SBR was operated in two cycles of operation with 12 hours per cycle for 187 days using biofilm formed on polyethylene carriers. Biofilm carriers were used to aim a higher specific surface area for the attached biofilm and for obtainability of anammox process. The seeding sludges were from landfill leachate treatment plant and reject water of municipal wastewater treatment plant in Germany. Synthetic water was used for the influent to be able to control the nutrient. Phase I was initial phase with temperature between 24-30°C, phase II was recovery phase after pH reduction shock occurred and phase III was operated with declining temperature to 5°C. During last phase, each half-day cycle was observed based on hourly measurements of inorganic nitrogen components for temperatures 25°C, 20°C, 15°C, 10°C, and 5°C in pursuance of the analyzing the effect of reduction temperature on NRE and microbial community structure. Dry matter tests, i.e. MLSS and MLVSS, was done to know the physical presence of microbial community based on biomass concentration, while FISH/qFISH and 16s rRNA high-throughput sequencing were used for identification of the microbial community structure. In phase I, the system could effectually remove $NH_4^+$-N during simultaneous nitrification, anammox and denitrification (SNAD). Results showed that NRE in the phase I was able to reach up to 99.92 % on day 28 based on $NH_4^+$-N concentration and 91.24 % based on total inorganic nitrogen (TIN) concentration. Whereas for nitrogen removal rate (NRR), the highest values are 0.14 kg m$^{-3}$ day$^{-1}$ based on

$NH_4^+$-N concentration with NRE 96.57 % and 0.15 kgN $m^{-3}$ $day^{-1}$ with NRE 83.6 % based on TIN concentration, both on the day 23. The intermittently regulated aeration pattern consequence the highest NRE with average dissolved oxygen (DO) concentration of 1.2 mg $L^{-1}$.

### 6.3.2 Adapting with low temperature

Reduction of temperature caused an expected declination of NRE, although a good adaptation was able to be achieved on 20°C. The highest of MLSS & MLVSS values were also occurred on the day 28, which means that a high-density population of bacteria on the fixed-film activated sludge contributed for higher NRE. AOB was major bacteria group inside the reactor on day 27 from 16s rRNA sequencing results based on relative abundancy and from qFISH results in phase III based on relative abundance of targeted microbial groups.

The IFAS is a hybrid variation of the moving bed biofilm reactor (MBBR) based process from the integration of biofilm carrier technology within conventional activated sludge. This technology uses synthetic packing materials, which will be suspended in the activated-sludge mixed liquor or fixed in the reactor. It intends to enhance the activated-sludge process by providing a greater biomass concentration in reactor, hence the potential to reduce the basin size requirements. It can also improve volumetric nitrification rates and accomplish denitrification and anammox process in the same reactor, where the aeration is given, by having anoxic zones within the biofilm depth (Figure 7.) (Tchobanoglous et al., 2003).

### 6.3.3 Why IFAS?

Anammox-based processes were successfully implemented at several full-scale plants to treat warm ammonium-rich reject water from anaerobic digesters (side stream) (van der Star et al., 2007; Wett, 2007). Due to the advantages of this technology in the side stream (1 % of the volume flow of the wastewater treatment plant, WWTP), current scientific research is focused on the implementation of anammox technology in the mainstream (99 % of the volume flow) of the WWTP (Kartal et al., 2010). However, different challenges need to be overcome for successful application of this process in the mainstream. As opposed to the side stream, the mainstream has lower temperature (down to ca. 10 °C instead of 30 °C), stronger effluent requirements, lower ammonium concentrations (ca. 50 gN $m^{-3}$ instead of 500 gN $m^{-3}$).

Conventional activated sludge bioreactors are retrofitted with the addition of IFAS carrier retaining screens and modifications to the aeration grid to accommodate the addition of the IFAS biofilm carriers. The media facilitates the growth of attached biomass and due to its size, is fluidized throughout the bioreactor. Carrier size, geometry and specific internal surface area are critical features. The upgrade to IFAS or MBBR often consists of simply adding carriers to existing basins and can therefore be completed in a cost-effective and timely manner without major civil engineering requirements and no requirement for additional land. PLC based control system optimize IFAS/MBBR process performance by minimizing energy and costs. A comparison of IFAS and MBBR is shown in Fig. 54.

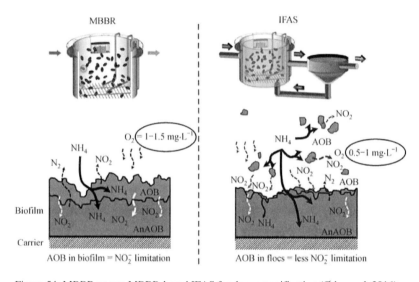

Figure 54: MBBR versus MBBR-based IFAS for de-ammonification (Ødegaard, 2016)

The main feature of IFAS process includes high surface area for microbial growth, which leads to enhanced nitrification rates along with being capable to bear organic as well as hydraulic shock loads. This system is categorized into two types of modules based on the arrangement of biocarriers inside the bioreactor, which are fixed packing and suspended packing. Fixed packing means the packing materials are fixed inside the reactor and can't move around freely. Whereas the suspended packing means the synthetic packing materials will be suspended in the reactor, which can still be moving around. The presence of carriers discourages the use of more efficient fine bubble aeration equipment, since this equipment requires periodic drainage of the aeration and removal of the packing for diffuser cleaning. The suspended packing type facilitates high oxygen and nutrient transfers in reactors (Ye et al., 2009), which is why this type was used in this experiment. Operating the mobile IFAS is better and more flexible in sequencing batch reactor (SBR) due to no clarifier is needed for SBR since the sedimentation will take place in the reactor itself (U.S.Environmental Protection Agency, 1999). IFAS-SBR combination could improve the nitrification efficiency of conventional SBR and reduced the capital cost of upgrading existing reaction tanks. However, it is important to understand how the bacterial community changes in response to different operational conditions. To effectively enrich the slow growing organisms, such as methanogen, anammox and nitrifier, granular sludge is another option (Nicolella et al., 2000). It provides a long SRT and has an internal anoxic zone, which could promote the growth of anammox bacteria (Winkler et al., 2012). A high removal loads of pollutants and high resistance to the influent fluctuation could be achieved by granular sludge system (Carvajal-Arroyo et al., 2013).

## 6.4 Future roadmap

Nowadays the road is through energy neutral WWTPs. As Ødegaard, (2016) showed, concepts for domestic wastewater treatment plants of the future are discussed. This concept is possible either using a) based on established, compact, proven technologies (i.e. nitrification-denitrification for N-removal in the mainstream and b) on emerging, compact technologies (i.e. deammonification in two steps together with final polishing with SAD or SNAD in on single reactor the main stream. The latter will give an energy neutral wastewater treatment plant, while this cannot be guaranteed for the first one. Both concepts can minimize energy consumption by using compact biological and physical/chemical processes combined in an optimal way, for instance by using moving bed biofilm reactor (MBBR) processes for biodegradation and high-rate particle separation processes, and deammonification processes for N-removal and maximize energy (biogas) production through digestion by using wastewater treatment processes that minimize biodegradation of the sludge (prior to digestion) and pre-treatment of the sludge prior to digestion by thermal hydrolysis.

To date, feasibility of mainstream PN/A process has been demonstrated and proven by experimental results at various scales although with the low loading rates and elevated nitrogen concentration in the effluent at low temperatures (15–10 °C) and still a lot of research and developments are undergoing.

## References

Adav, S. S., Lee, D., Show, K., & Tay, J. (2008). Aerobic granular sludge : Recent advances, *i*, 411–423. https://doi.org/10.1016/j.biotechadv.2008.05.002

Akunna, J.C., Bizeau, C., Moletta, R., 1993. Nitrate and nitrite reductions with anaerobic sludge using various carbon sources: Glucose, glycerol, acetic acid, lactic acid and methanol. Water Res. 27, 1303–1312. doi:10.1016/0043-1354(93)90217-6

Ali, M., Okabe, S., 2015. Anammox-based technologies for nitrogen removal: Advances in process start-up and remaining issues. Chemosphere 141, 144–153. doi:10.1016/j.chemosphere.2015.06.094

Ali, M., Oshiki, M., Awata, T., Isobe, K., Kimura, Z., Yoshikawa, H., Hira, D., Kindaichi, T., Satoh, H., Fujii, T., Okabe, S., 2015. Physiological characterization of anaerobic ammonium oxidizing bacterium "CandidatusJettenia caeni." Environ. Microbiol. 17, 2172–2189. doi:10.1111/1462-2920.12674

Almstrand, R., Persson, F., Daims, H., Ekenberg, M., Christensson, M., Wilén, B.M., Sörensson, F., Hermansson, M., 2014. Three-dimensional stratification of bacterial biofilm populations in a moving bed biofilm reactor for nitritation-anammox. Int. J. Mol. Sci. 15, 2191–2206. doi:10.3390/ijms15022191

Almstrand, R., Persson, F., Daims, H., Ekenberg, M., Christensson, M., Wilén, B.M., Sörensson, F. & Hermansson, M. 2014. Three-dimensional stratification of bacterial biofilm populations in a moving bed biofilm reactor for nitritation-anammox. Int. J. Mol. Sci. 15, 2191–2206. doi:10.3390/ijms15022191

Anjali, G., & Sabumon, P. C. 2017. Development of simultaneous partial nitrification, anammox and denitrification (SNAD) in a non-aerated SBR. *International Biodeterioration and Biodegradation*, *119*(2), 43–55. https://doi.org/10.1016/j.ibiod.2016.10.047

Anthonisen, A. C., Loehr, R. C., Prakasam, T. B. S., & Srinath, E. G. 1976. Inhibition of Nitrification by Ammonia and Nitrous Acid. *Journal (Water Pollution Control Federation)*, *48*(5), 835–852. Retrieved from http://www.jstor.org/stable/25038971

Arashiro, L.T., Rada-Ariza, A.M., Wang, M., Van Der Steen, P. and Ergas, S.J., 2017. Modeling shortcut nitrogen removal from wastewater using an algal–bacterial consortium. Water Science and Technology, 75(4), pp.782-792.

Awata, T., Oshiki, M., Kindaichi, T., Ozaki, N., Ohashi, A., Okabe, S., 2013. Physiological characterization of an anaerobic ammonium-oxidizing bacterium belonging to the Ca. Scalindua group. Appl Environ Microb. 79(13), 4145-4148.

## References

Azari, M., Walter, U., Rekers, V., Gu, J.D. & Denecke, M. 2017a. More than a decade of experience of landfill leachate treatment with a full-scale anammox plant combining activated sludge and activated carbon biofilm. Chemosphere, 174, 117-126.

Azari, M., Le, A.V. and Denecke, M., 2017b, Population dynamic of microbial consortia in a granular activated carbon-Assisted biofilm reactor: lessons from modeling. In Frontiers International Conference on Wastewater Treatment and Modeling (pp. 588-595). Springer, Cham.

Azari, M., Lübken, M. & Denecke, M. 2017c. Simulation of simultaneous anammox and denitrification for kinetic and physiological characterization of microbial community in a granular biofilm system. Biochemical Engineering Journal, 127, pp.206-216.

Azari, M., Le, A.V., Lübken, M. and Denecke, M., 2018. Model-based analysis of microbial consortia and microbial products in an anammox biofilm reactor. Water Science and Technology, p.wst2018081.

Barker, D.J., Stuckey, D.C., 1999. A review of soluble microbial products (SMP) in wastewater treatment systems. Water Res. doi:10.1016/S0043-1354(99)00022-6

Beun, J. J., Heijnen, J. J., & Loosdrecht, M. C. M. Van. 2001. N-Removal in a Granular Sludge Sequencing Batch Airlift Reactor N-Removal in a Granular Sludge, (OCTOBER 2001). https://doi.org/10.1002/bit.1167

Beun, J. J., Hendriks, A., Van Loosdrecht, M. C. M., Morgenroth, E., Wilderer, P. A., & Heijnen, J. J. 1999. Aerobic granulation in a sequencing batch reactor. *Water Research*, *33*(10), 2283–2290. https://doi.org/10.1016/S0043-1354(98)00463-1

Boltz, J.P., Smets, B.F., Rittmann, B.E., van Loosdrecht, M.C.M., Morgenroth, E., Daigger, G.T., 2017. From biofilm ecology to reactors: a focused review. Water Sci. Technol. wst2017061. doi:10.2166/wst.2017.061

Bothe, H., Jost, G., Schloter, M., Ward, B. B., & Witzel, K.-P. (2000). Molecular analysis of ammonia oxidation and denitrification in natural environments. *FEMS Microbiology Reviews*, *24*(5), 673–690.

Bouchez, T., Patureau, D., Dabert, P., Juretschko, S., Doré, J., Delgenès, P., Moletta, R., Wagner, M., 2000. Ecological study of a bioaugmentation failure. Environ. Microbiol. 2, 179–190. doi:10.1046/j.1462-2920.2000.00091.x

Brouwer, M., Van Loosdrecht, M. C. M., & Heijnen, J. J. 1996. One reactor system for ammonium removal via nitrite. *STOWA Report*, 1–96.

Cao, S., Du, R., Niu, M., Li, B., Ren, N. and Peng, Y., 2016. Integrated anaerobic ammonium oxidization with partial denitrification process for advanced nitrogen removal from high-strength wastewater. Bioresource technology, 221, pp.37-46.

Carvajal-Arroyo, J. M., Sun, W., Sierra-Alvarez, R., & Field, J. A. 2013. Inhibition of anaerobic ammonium oxidizing (anammox) enrichment cultures by substrates, metabolites and common wastewater constituents. *Chemosphere*, *91*(1), 22–27.

Castro-Barros, C.M., Jia, M., van Loosdrecht, M.C., Volcke, E.I. and Winkler, M.K., 2017. Evaluating the potential for dissimilatory nitrate reduction by anammox bacteria for municipal wastewater treatment. Bioresource technology, 233, pp.363-372.

Chrispim, M.C. and Nolasco, M.A., 2017. Greywater treatment using a moving bed biofilm reactor at a university campus in Brazil. Journal of Cleaner Production, 142, pp.290-296.

Christensson, M., Ekström, S., Chan, A.A., Le Vaillant, E., Lemaire, R., 2013. Experience from start-ups of the first ANITA Mox plants. Water Sci Technol. 67(12), 2677-2684.

Chu, Z., Wang, K., Li, X., Zhu, M., Yang, L., Zhang, J., 2015. Microbial characterization of aggregates within a one-stage nitritation–anammox system using high-throughput amplicon sequencing. Chem. Eng. J. 262, 41–48.

Cirpus, I. E. Y., de Been, M., Op den Camp, H. J. M., Strous, M., Le Paslier, D., Kuenen, G. J., & Jetten, M. S. M. (2005). A new soluble 10 kDa monoheme cytochrome c-552 from the anammox bacterium Candidatus"Kuenenia stuttgartiensis." *FEMS Microbiology Letters*, *252*(2), 273–278. Retrieved from http://dx.doi.org/10.1016/j.femsle.2005.09.007

Clesceri, L. S., Eaton, A. D., Greenberg, A. E., Association, A. P. H., Association, A. W. W., & Federation, W. E. (1998). *Standard Methods for the Examination of Water and Wastewater*. American Public Health Association. Retrieved from https://books.google.de/books?id=pv5PAQAAIAAJ

Code, T.U.S., Chao, Y., Mao, Y., Yu, K., Zhang, T., 2016. Notice : This material may be protected by Copyright Law Novel nitrifiers and comammox in a full-scale hybrid biofilm and activated sludge reactor revealed by metagenomic approach 8225–8237. doi:10.1007/s00253-016-7655-9

Coughlan, L.M., Cotter, P.D., Hill, C. and Alvarez-Ordóñez, A., 2016. New weapons to fight old enemies: novel strategies for the (bio) control of bacterial biofilms in the food industry. Frontiers in microbiology, 7, p.1641.

Daims, H., Brühl, A., Amann, R., Schleifer, K.H. and Wagner, M., 1999. The domain-specific probe EUB338 is insufficient for the detection of all Bacteria: development and evaluation of a more comprehensive probe set. Syst Appl Micriobiol. 22(3), 434-444.

## References

Daims, H., L??cker, S., Wagner, M., 2016. A New Perspective on Microbes Formerly Known as Nitrite-Oxidizing Bacteria. Trends Microbiol. doi:10.1016/j.tim.2016.05.004

Daims, H., Lebedeva, E.V., Pjevac, P., Han, P., Herbold, C., Albertsen, M., Jehmlich, N., Palatinszky, M., Vierheilig, J., Bulaev, A., Kirkegaard, R.H., von Bergen, M., Rattei, T., Bendinger, B., Nielsen, P.H., Wagner, M., 2015. Complete nitrification by Nitrospira bacteria. Nature 528, 504–509. doi:10.1038/nature16461

Daims, H., Lücker, S., Wagner, M., 2006. Daime, a Novel Image Analysis Program for Microbial Ecology and Biofilm Research. Environ. Microbiol. 8, 200–213. doi:10.1111/j.1462-2920.2005.00880.x

Daims, H., Nielsen, P.H., Nielsen, J.L., Juretschko, S., Wagner, M., 2000. Novel Nitrospira-like bacteria as dominant nitrite-oxidizers in biofilms from wastewater treatment plants: diversity and in situ physiology. Water Sci. Technol. 41, 85–90.

Daims, H., Wagner, M., 2007. Quantification of uncultured microorganisms by fluorescence microscopy and digital image analysis. Appl. Microbiol. Biotechnol. 75, 237–248. doi:10.1007/s00253-007-0886-z

Dapena-Mora, A., Van Hulle, S.W.H., Campos, J.L., Méndez, R., Vanrolleghem, P.A., Jetten, M., 2004. Enrichment of Anammox biomass from municipal activated sludge: Experimental and modeling results. J. Chem. Technol. Biotechnol. 79, 1421–1428. doi:10.1002/jctb.1148

Daverey, A., Chen, Y.C., Dutta, K., Huang, Y.T., Lin, J.G., 2015. Start-up of simultaneous partial nitrification, anammox and denitrification (SNAD) process in sequencing batch biofilm reactor using novel biomass carriers. Bioresource Technol. 190, pp.480-486

Dosta, J., Fernández, I., Vázquez-Padín, J. R., Mosquera-Corral, A., Campos, J. L., Mata-Álvarez, J., & Méndez, R. 2008. Short- and long-term effects of temperature on the Anammox process. *Journal of Hazardous Materials*, *154*(1–3), 688–693. https://doi.org/10.1016/j.jhazmat.2007.10.082

Dunkel, T., de León Gallegos, E. L., Schönsee, C. D., Hesse, T., Jochmann, M., Wingender, J., & Denecke, M. 2016. Evaluating the influence of wastewater composition on the growth of Microthrix parvicella by GCxGC/qMS and real-time PCR. *Water Research*, *88*, 510–523. https://doi.org/10.1016/j.watres.2015.10.027

Egli, K., Fanger, U., Alvarez, P. J. J., Siegrist, H., Van der Meer, J. R., & Zehnder, A. J. B. 2001. Enrichment and characterization of an anammox bacterium from a rotating biological contactor treating ammonium-rich leachate. *Archives of Microbiology*, *175*(3), 198–207. https://doi.org/10.1007/s002030100255

ElMekawy, A., Mohanakrishna, G., Srikanth, S. and Pant, D., 2016. The Role of Bioreactors in Industrial Wastewater Treatment. Environmental Waste Management, p.157.

Federation, W. E. 2010. *Biofilm Reactors WEF MOP 35*. McGraw-Hill Education. Retrieved from https://books.google.de/books?id=LaniXghYy9MC

Fernández, I., Vázquez-Padín, J.R., Mosquera-Corral, A., Campos, J.L. and Méndez, R., 2008. Biofilm and granular systems to improve Anammox biomass retention. Biochemical Engineering Journal, 42(3), pp.308-313.

Flemming, H.-C., Szewzyk, U., Griebe, T., 2000. Biofilms: investigative methods and applications. CRC Press.

Flemming, H.C., Wingender, J., 2010. The biofilm matrix. Nat Rev Microbiol. 8(9), 623-633.

Fux, C., Boehler, M., Huber, P., Brunner, I., & Siegrist, H. (2002). Biological treatment of ammonium-rich wastewater by partial nitritation and subsequent anaerobic ammonium oxidation (anammox) in a pilot plant. *Journal of Biotechnology*, *99*(3), 295–306. https://doi.org/10.1016/S0168-1656(02)00220-1

Gayle, B. P., Boardman, G. D., Sherrard, J. H., & Benoit, R. E. (1989). Biological denitrification of water. *Journal of Environmental Engineering*, *115*(5), 930–943.

Gerardi, M. H. 2003. *Nitrification and Denitrification in the Activated Sludge Process*. Wiley. Retrieved from https://books.google.de/books?id=ccIKuXYNlBMC

Gilbert, E. M., Agrawal, S., Karst, S. M., Horn, H., Nielsen, P. H., & Lackner, S. 2014. Low temperature partial nitritation/anammox in a moving bed biofilm reactor treating low strength wastewater. *Environmental Science and Technology*, *48*(15), 8784–8792. https://doi.org/10.1021/es501649m

Grady, C.P.L., Daigger, G.T., Lim, H.C., 1999. Biological wastewater treatment. Hazard. Waste October, 1076.

Gujer, W., Henze, M., Mino, T., Vanloosdrecht, M., 1999. Activated Sludge Model No. 3. Water Sci. Technol. 39, 183–193. doi:10.1016/S0273-1223(98)00785-9

Gujer, W., Jenkins, D., 1975. The contact stabilization activated sludge process-oxygen utilization, sludge production and efficiency. Water Res. 9, 553–560. doi:10.1016/0043-1354(75)90081-0

Han, P., Gu, J.D., 2015. Further analysis of anammox bacterial community structures along an anthropogenic nitrogen-input gradient from the riparian sediments of the Pearl River Delta to the deep-ocean sediments of the South China Sea. Geomicrobiol J. 32 (9), 789⊓798.

# References

Heitz, E., Flemming, H.-C., Sand, W., 1997. Microbially influenced corrosion of materials: scientific and engineering aspects. Solid State Electrochem 1. doi:3540604324

Henze, M., Gujer, W., Mino, T., Matsuo, T., Wentzel, M.C., Marais, G., 1995. Activated Sludge Model No. 2, IAWQ Scientific and Technical reports, No. 3, IAWQ, London. ISBN 1, 0.

Hidaka, T., Yamada, H., Kawamura, M., & Tsuno, H. 2002. Effect of dissolved oxygen conditions on nitrogen removal in continuously fed intermittent-aeration process with two tanks. *Water Science and Technology*, *45*(12), 181 LP-188. Retrieved from http://wst.iwaponline.com/content/45/12/181.abstract

Hong, Y.G., Li, M., Cao, H., Gu, J.D., 2011. Residence of habitat-specific anammox bacteria in the deep-sea subsurface sediments of the South China Sea: analyses of marker gene abundance with physical chemical parameters. Microb Ecol. 62(1), 36-47.

Horn, H., Lackner, S., 2014. Modeling of biofilm systems: A review. Adv. Biochem. Eng. Biotechnol. 146, 53–76. doi:10.1007/10_2014_275

Wickens, M. and Cox, M.M., 2009. Critical reviews in biochemistry and molecular biology. Introduction. Critical reviews in biochemistry and molecular biology, 44(1), pp.2-2.

Jetten, M.S., Strous, M., Van de Pas-Schoonen, K.T., Schalk, J., van Dongen, U.G., van de Graaf, A.A., Logemann, S., Muyzer, G., van Loosdrecht, M.C. and Kuenen, J.G., 1998. The anaerobic oxidation of ammonium. FEMS Microbiology reviews, 22(5), pp.421-437.

Jetten, M. S. M., Wagner, M., Fuerst, J., Van Loosdrecht, M., Kuenen, G., & Strous, M. 2001. Microbiology and application of the anaerobic ammonium oxidation ('anammox') process. *Current Opinion in Biotechnology*, *12*(3), 283–288. https://doi.org/10.1016/S0958-1669(00)00211-1

Jetten, M.S.M., Schmid, M., Schmidt, I., Wubben, M., van Dongen, U., Abma, W., Sliekers, O., Revsbech, N.P., Beaumont, H.J.E., Ottosen, L., Volcke, E., Laanbroek, H.J., Campos-Gomez, J.L., Cole, J., van Loosdrecht, M., Mulder, J.W., Fuerst, J., Richardson, D., van de Pas, K., Mendez-Pampin, R., Third, K., Cirpus, I., van Spanning, R., Bollmann, A., Nielsen, L.P., den Camp, H.O., Schultz, C., Gundersen, J., Vanrolleghem, P., Strous, M., Wagner, M., Kuenen, J.G., 2002. Improved nitrogen removal by application of new nitrogen-cycle bacteria. Rev. Environ. Sci. Biotechnol. doi:10.1023/A:1015191724542

Jetten, M.S.M., Strous, M., Van de Pas-Schoonen, K.T., Schalk, J., van Dongen, U.G.J.M., van de Graaf, A.A., Logemann, S., Muyzer, G., van Loosdrecht, M.C.M., Kuenen, J.G.,

1998. The anaerobic oxidation of ammonium. FEMS Microbiol. Rev. 22, 421–437.

Jin, R.C., Yang, G.F., Yu, J.J., Zheng, P., 2012. The inhibition of the Anammox process: A review. Chem. Eng. J. 197, 67–79. doi:10.1016/j.cej.2012.05.014

Jonasson, M. and Ulf Jeppsson, I.E.A., 2007. Energy Benchmark for Wastewater Treatment Processes (Doctoral dissertation, MS Thesis, 2007, Dept. of Industrial Electrical Engineering and Automation Lund University).

Juretschko, S., Loy, A., Lehner, A., Wagner, M., 2002. The microbial mommunity momposition of a nitrifying-denitrifying activated sludge from an industrial sewage treatment plant analyzed by the full-cycle rRNA approach. Syst. Appl. Microbiol. 25, 84–99. doi:10.1078/0723-2020-00093

Rosenwinkel, K.H. and Cornelius, A., 2005. Deammonification in the moving-bed process for the treatment of wastewater with high ammonia content. Chemical engineering & technology, 28(1), pp.49-52.

Kartal, B., Koleva, M., Arsov, R., van der Star, W., Jetten, M.S.M., Strous, M., 2006. Adaptation of a freshwater anammox population to high salinity wastewater. J. Biotechnol. 126, 546–553.

Kartal, B., Kuenen, J.V. and Van Loosdrecht, M.C.M., 2010. Sewage treatment with anammox. Science, 328(5979), pp.702-703.

Kartal, B., Kuypers, M.M., Lavik, G., Schalk, J., Op den Camp, H.J., Jetten, M.S. and Strous, M., 2007. Anammox bacteria disguised as denitrifiers: nitrate reduction to dinitrogen gas via nitrite and ammonium. Environmental microbiology, 9(3), pp.635-642.

Ke, Y., 2014. Simultaneous Anammox and Denitrification (SAD) Process with Anammox Granular Sludge. Shaker Verlag GmbH. Germany.

Ke, Y., Azari, M., Han, P., Görtz, I., Gu, J.D. and Denecke, M., 2015. Microbial community of nitrogen-converting bacteria in anammox granular sludge. International Biodeterioration & Biodegradation, 103, pp.105-115.

Khramenkov, S.V., Kozlov, M.N., Kevbrina, M.V., Dorofeev, A.G., Kazakova, E.A., Grachev, V.A., Kuznetsov, B.B., Polyakov, D.Y., Nikolaev, Y.A., 2013. A novel bacterium carrying out anaerobic ammonium oxidation in a reactor for biological treatment of the filtrate of wastewater fermented sludge. Microbiology. 82(5), 628-636.

Khin, T. and Annachhatre, A.P., 2004. Novel microbial nitrogen removal processes. Biotechnology advances, 22(7), pp.519-532.

Kim, H. su, Gellner, J. W., Boltz, J. P., Freudenberg, R. G., Gunsch, C. K., & Schuler, A. J. (2010). Effects of integrated fixed film activated sludge media on activated sludge settling in biological nutrient removal systems. *Water Research*, 44(5), 1553–1561.

https://doi.org/10.1016/j.watres.2009.11.001

Kindaichi, T., Awata, T., Mugimoto, Y., Rathnayake, R.M., Kasahara, S., Satoh, H., 2016. Effects of organic matter in livestock manure digester liquid on microbial community structure and in situ activity of anammox granules. Chemosphere. 159, pp.300-307.

Kindaichi, T., Ito, T., Okabe, S., 2004. Ecophysiological Interaction between Nitrifying Bacteria and Heterotrophic Bacteria in Autotrophic Nitrifying Biofilms as Determined by Microautoradiography-Fluorescence In Situ Hybridization. Appl. Environ. Microbiol. 70, 1641–1650. doi:10.1128/AEM.70.3.1641-1650.2004

Kindaichi, T., Tsushima, I., Ogasawara, Y., Shimokawa, M., Ozaki, N., Satoh, H., Okabe, S., 2007. In situ activity and spatial organization of anaerobic ammonium-oxidizing (anammox) bacteria in biofilms. Appl Environ Microb. 73(15), 4931-4939.

Klimenko, A.I., Mustafin, Z.S., Chekantsev, A.D., Zudin, R.K., Matushkin, Y.G. and Lashin, S.A., 2016. A review of simulation and modeling approaches in microbiology. Russian Journal of Genetics: Applied Research, 6(8), pp.845-853.

Kosari, S.F., Rezania, B., Lo, K.V., Mavinic, D.S., 2014. Operational strategy for nitrogen removal from centrate in a two-stage partial nitrification–Anammox process. Environ. Technol. 35(9), 1110-1120

Kovarova-Kovar, K., Egli, T., 1998. Growth Kinetics of Suspended Microbial Cells: From Single-Substrate-Controlled Growth to Mixed-Substrate Kinetics. Microbiol. Mol. Biol. Rev. 62, 646–666.

Kuenen, J.G., 2008. Anammox bacteria: from discovery to application. Nature Reviews Microbiology, 6(4), p.320.

Kuypers, M. M. M., Marchant, H. K., & Kartal, B. 2018. The microbial nitrogen-cycling network. *Nature Reviews Microbiology*.

Lackner, S., Gilbert, E.M., Vlaeminck, S.E., Joss, A., Horn, H., van Loosdrecht, M.C.M., 2014. Full-scale partial nitritation/anammox experiences - An application survey. Water Res. 55, 292–303. doi:10.1016/j.watres.2014.02.032

Lackner, S., Thoma, K., Gilbert, E.M., Gander, W., Schreff, D., Horn, H., 2015. Start-up of a full-scale deammonification SBR-treating effluent from digested sludge dewatering. Water Sci Technol. 71(4), 553-559.

Lan, C.J., Kumar, M., Wang, C.C. and Lin, J.G., 2011. Development of simultaneous partial nitrification, anammox and denitrification (SNAD) process in a sequential batch reactor. Bioresource Technology, 102(9), pp.5514-5519.

## References

Langone, M., 2013. Simultaneous partial nitritation, Anammox and denitrification (SNAD) process for treating ammonium-rich wastewaters (Doctoral dissertation, University of Trento).

Lariyah, M.S., Mohiyaden, H.A., Hayder, G., Hussein, A., Basri, H., Sabri, A.F. and Noh, M.N., 2016, March. Application of moving bed biofilm reactor (MBBR) and integrated fixed activated sludge (IFAS) for biological river water purification system: a short review. In IOP Conference Series: Earth and Environmental Science (Vol. 32, No. 1, p. 012005). IOP Publishing.

Laspidou, C.S., Rittmann, B.E., 2002a. A unified theory for extracellular polymeric substances, soluble microbial products, and active and inert bi omass. Water Res. 36, 2711–2720. doi:10.1016/S0043-1354(01)00413-4

Laspidou, C.S., Rittmann, B.E., 2002b. Non-steady state modeling of extracellular polymeric substances, soluble microbial products, and active and inert biomass. Water Res. 36, 1983–1992. doi:10.1016/S0043-1354(01)00414-6

Lee, K., Song, K., Snd, D., Nitrite, V. I. A., & An, I. N. 1999. Nitrogen removal from synthetic wastewater by simultaneous nitrification and denitrification ( SND ) via nitrite in an ..., 33(February 2014). https://doi.org/10.1016/S0043-1354(98)00159-6

Lee, K.H., Wang, Y.F., Zhang, G.X.,Gu, J.D., 2014. Distribution patterns of ammonia-oxidizing bacteria and anammox bacteria in the freshwater marsh of Honghe wetland in Northeast China. Ecotoxicology. 23(10) 1930-1942.

Lee, K.H., Wang, Y.F., Wang, Y., Gu, J.D., Jiao, J.J., 2016. Abundance and diversity of aerobic/anaerobic ammonia/ammonium-oxidizing microorganisms in an ammonium-rich aquitard in the pearl river delta of south China. Microb Ecol. 1-11.

Li, M., Cao, H., Hong, Y., Gu, J.D., 2013. Using the variation of anammox bacteria community structures as a bio-indicator for anthropogenic/terrestrial nitrogen inputs in the Pearl River Delta (PRD). Appl Microbiol Biot. 97(22), 9875-9883.

Li, J., Elliott, D., Nielsen, M., Healy, M. G., & Zhan, X. 2011. Long-term partial nitrification in an intermittently aerated sequencing batch reactor ( SBR ) treating ammonium-rich wastewater under controlled oxygen-limited conditions. *Biochemical Engineering Journal*, 55(3), 215–222. https://doi.org/10.1016/j.bej.2011.05.002

Li, J., Qiang, Z., Yu, D., Wang, D., Zhang, P. and Li, Y., 2016. Performance and microbial community of simultaneous anammox and denitrification (SAD) process in a sequencing batch reactor. Bioresource technology, 218, pp.1064-1072.

Lindsay, M.R., Webb, R.I., Strous, M., Jetten, M.S.M., Butler, M.K., Forde, R.J., Fuerst,

J.A., 2001. Cell compartmentalisation in planctomycetes: Novel types of structural organisation for the bacterial cell. Arch. Microbiol. 175, 413–429. doi:10.1007/s002030100280

Liu, T., Ma, B., Chen, X., Ni, B.J., Peng, Y. and Guo, J., 2017. Evaluation of mainstream nitrogen removal by simultaneous partial nitrification, anammox and denitrification (SNAD) process in a granule-based reactor. Chemical Engineering Journal, 327, pp.973-981.

Liu, Y., Sun, J., Peng, L., Wang, D., Dai, X., Ni, B.-J., 2016. Assessment of Heterotrophic Growth Supported by Soluble Microbial Products in Anammox Biofilm using Multidimensional Modeling. Sci. Rep. 6, 27576. doi:10.1038/srep27576

Lotti, T., Kleerebezem, R., & van Loosdrecht, M. C. M. (2015). Effect of temperature change on anammox activity. *Biotechnology and Bioengineering, 112*(1), 98–103. https://doi.org/10.1002/bit.25333

Lotti, T., Kleerebezem, R., Lubello, C. and Van Loosdrecht, M.C.M., 2014. Physiological and kinetic characterization of a suspended cell anammox culture. Water research, 60, pp.1-14.

Lu, H.F., Zheng, P., Ji, Q.X., Zhang, H.T., Ji, J.Y., Wang, L., Ding, S., Chen, T.T., Zhang, J.Q., Tang, C.J., Chen, J.W., 2012. The structure, density and settlability of anammox granular sludge in high-rate reactors. Bioresource Technol. 123, 312-317.

Ma, B., Qian, W., Yuan, C., Yuan, Z. & Peng, Y. 2017. Achieving mainstream nitrogen removal through coupling anammox with denitratation. Environmental science & technology, 51(15), pp.8405-8413.

Ma, B., Wang, S., Cao, S., Miao, Y., Jia, F., Du, R. and Peng, Y., 2016. Biological nitrogen removal from sewage via anammox: recent advances. Bioresource technology, 200, pp.981-990.

Madigan, M.T., Martinko, J.M., Stahl, D.A., Clark, D.P., 2012. Brock Biology of Microorganisms, 13th Edition, International Microbiology. doi:10.1038/hr.2014.17

McCarty, P.L., Beck, L., St Amant, P., 1969. Biological denitrification of agricultural wastewaters by addition of organic materials. Purdue Univ Eng Ext Ser.

McSwain, B.S., Irvine, R.L., Hausner, M., Wilderer, P.A., 2005. Composition and distribution of extracellular polymeric substances in aerobic flocs and granular sludge. Appl Environ Microb. 71(2), 1051-1057.

Mesquita, P. da L., Aquino, S.F. de, Xavier, A.L.P., Silva, J.C., Afonso, R.C.F., Silva, S.Q., 2010. Soluble microbial product (SMP) characterization in bench-scale aerobic and anaerobic CSTRs under different operational conditions. Brazilian J. Chem. Eng. 27,

101–111.

Mobarry, B.K., Wagner, M., Urbain, V., Rittmann, B.E. & Stahl, D.A. 1997. Phylogenetic probes for analyzing abundance and spatial organization of nitrifying bacteria. Appl. Environ. Microbiol. 63, 815.Namkung, E. and Rittmann, B.E., 1986. Soluble microbial products (SMP) formation kinetics by biofilms. Water Research, 20(6), pp.795-806.

Mobarry, B.K., Wagner, M., Urbain, V., Rittmann, B.E., Stahl, D.A., 1997. Phylogenetic probes for analyzing abundance and spatial organization of nitrifying bacteria. Appl. Environ. Microbiol. 63, 815.

Mozumder, M.S.I., Picioreanu, C., van Loosdrecht, M.C. and Volcke, E.I., 2014. Effect of heterotrophic growth on autotrophic nitrogen removal in a granular sludge reactor. Environmental technology, 35(8), pp.1027-1037.

Mulder, A., van de Graaf, A.A., Robertson, L.A., Kuenen, J.G., 1995. Anaerobic ammonium oxidation discovered in a denitrifying fluidized bed reactor. FEMS Microbiol. Ecol. 16, 177–183. doi:10.1016/0168-6496(94)00081-7

Namkung, E., Rittmann, B.E., 1986. Soluble microbial products (SMP) formation kinetics by biofilms. Water Res. 20, 795–806. doi:10.1016/0043-1354(86)90106-5

Nelson, L. M., & Knowles, R. 1978. Effect of oxygen and nitrate on nitrogen fixation and denitrification by Azospirillum brasilense grown in continuous culture. *Canadian Journal of Microbiology*, 24(11), 1395–1403.

Ni, B.J., Chen, Y.P., Liu, S.Y., Fang, F., Xie, W.M. and Yu, H.Q., 2009. Modeling a granule-based anaerobic ammonium oxidizing (ANAMMOX) process. Biotechnology and bioengineering, 103(3), pp.490-499.

Ni, B.J., Joss, A., Yuan, Z., 2014. Modeling nitrogen removal with partial nitritation and anammox in one floc-based sequencing batch reactor. Water Res. 67, 321–329. doi:10.1016/j.watres.2014.09.028

Ni, B.J., Ruscalleda, M., Smets, B.F., 2012. Evaluation on the microbial interactions of anaerobic ammonium oxidizers and heterotrophs in Anammox biofilm. Water Res. 46, 4645–4652. doi:10.1016/j.watres.2012.06.016

Ni, B.J., Xie, W.M., Chen, Y.P., Fang, F., Liu, S.Y., Ren, T.T., Sheng, G.P., Yu, H.Q., Liu, G. and Tian, Y.C., 2011. Heterotrophs grown on the soluble microbial products (SMP) released by autotrophs are responsible for the nitrogen loss in nitrifying granular sludge. Biotechnology and bioengineering, 108(12), pp.2844-2852.

Ni, S.Q. and Meng, J., 2011. Performance and inhibition recovery of anammox reactors seeded with different types of sludge. Water Science and Technology, 63(4), pp.710-718.

Ni, S.Q. and Yang, N., 2014. Evaluation of granular anaerobic ammonium oxidation process for the disposal of pre-treated swine manure. PeerJ, 2, p.e336.

Niu, Q., Zhang, Y., Ma, H., He, S., Li, Y.Y., 2016. Reactor kinetics evaluation and performance investigation of a long-term operated UASB-anammox mixed culture process. Int Biodetr Biodegr. 108, 24-33.

Nicolella, C., van Loosdrecht, M. C. M., & Heijnen, S. J. 2000. Particle-based biofilm reactor technology. *Trends in Biotechnology, 18*(7), 312–320.

Nielsen, P.H., Jahn, A., 1999. Extraction of EPS, in: Microbial Extracellular Polymeric Substances. Springer, pp. 49–72.

Nielsen, P.H., Lemmer, H., Daims, H., 2009. FISH Handbook for Biological Wastewater Treatment, Current.

Nogueira, R., Elenter, D., Brito, A., Melo, L.F., Wagner, M., Morgenroth, E., 2005. Evaluating heterotrophic growth in a nitrifying biofilm reactor using fluorescence in situ hybridization and mathematical modeling, Water Science and Technology.

Ødegaard, H., 2016. A road-map for energy-neutral wastewater treatment plants of the future based on compact technologies (including MBBR). Frontiers of Environmental Science & Engineering, 10(4), p.2.

Ostace, G.S., Cristea, V.M., Agachi, P.S., 2011. Cost reduction of the wastewater treatment plant operation by MPC based on modified ASM1 with two-step nitrification/denitrification model. Comput. Chem. Eng. 35, 2469–2479. doi:DOI:10.1016/j.compchemeng.2011.03.031

Parkin, G.F., McCarty, P.L., 1975. Characteristics and removal of soluble organic nitrogen in treated effluents. Prog. water Technol.

Per Halkjær Nielsen, H. D. and H. L. (2012). *FISH Handbook for Biological Wastewater Treatment. Uma ética para quantos?* (Vol. XXXIII). https://doi.org/10.1007/s13398-014-0173-7.2

Rahimi, Y., Torabian, A., Mehrdadi, N. and Shahmoradi, B., 2011. Simultaneous nitrification–denitrification and phosphorus removal in a fixed bed sequencing batch reactor (FBSBR). Journal of Hazardous Materials, 185(2-3), pp.852-857.

Randall, C. W., & Sen, D. 1996. Full-scale evaluation of an integrated fixed-film activated sludge (IFAS) process for enhanced nitrogen removal. *Water Science and Technology, 33*(12), 155–162. Retrieved from http://wst.iwaponline.com/content/33/12/155

Ranjbar, F., & Jalali, M. 2015. The effect of chemical and organic amendments on sodium

Reichert, P., 1994. Aquasim - A tool for simulation and data analysis of aquatic systems, in: Water Science and Technology. pp. 21–30.

Reichert, P., 1998. AQUASIM 2.0—user manual. Swiss Fed. Inst. Environ. Sci. Technol. Dubendorf, Switz.

Regmi, P., Holgate, B., Miller, M.W., Bunce, R., Park, H., Chandran, K., Wett, B., Murthy, S., Bott, C., 2013. NOB out-selection in mainstream makes two-stage deammonification and nitrite-shunt possible. In Proceedings of the Nutrient Removal and Recovery Trends in Resource Recovery and Use Conference. 28-31.

Rekers, V., Walter, U., Denecke, M., 2007. Method And Apparatus For Treating Wastewater With High Nitrogen And Low BOD 5 Share, Especially From Landfill Water,Patent No. EP 067 750 B1 2, Germany.

Reeve, P.J., Mouilleron, I., Chuang, H.P., Thwaites, B., Hyde, K., Dinesh, N., Krampe, J., Lin, T.F., van den Akker, B., 2016. Effect of feed starvation on side-stream anammox activity and key microbial populations. J Environ Manage. 171, pp.121-127.

Rittmann, B.E., Bae, W., Namkung, E., Lu, C.J., 1987. A critical evaluation of microbial product formation in biological processes. Water Sci. Technol. 19, 517–528.

Rittmann, B.E., Schwarz, A.O., Eberl, H.J., Morgenroth, E., Perez, J., van Loosdrecht, M., Wanner, O., 2004. Results from the multi-species benchmark problem (BM3) using one-dimensional models. Water Sci. Technol. 49, 163–168.

Rosselli, R., Romoli, O., Vitulo, N., Vezzi, A., Campanaro, S., de Pascale, F., Schiavon, R., Tiarca, M., Poletto, F., Concheri, G., 2016. Direct 16S rRNA-seq from bacterial communities: a PCR-independent approach to simultaneously assess microbial diversity and functional activity potential of each taxon. Sci. Rep. 6.

Rousseau, D., Verdonck, F., Moerman, O., Carrette, R., Thoeye, C., Meirlaen, J., Vanrolleghem, P.A., 2001. Development of a risk assessment based technique for design/retrofitting of WWTPs. Water Sci Technol. 43(7), 287-294.

Rusten, B., Hellström, B. G., Hellström, F., Sehested, O., Skjelfoss, E., & Svendsen, B. 2000. Pilot testing and preliminary design of moving bed biofilm reactors for nitrogen removal at the FREVAR wastewater treatment plant. *Water Science and Technology*, *41*(4–5), 13 LP-20. Retrieved from http://wst.iwaponline.com/content/41/4-5/13.abstract

Schmid, M., Walsh, K., Webb, R., Rijpstra, W.I.C., van de Pas-Schoonen, K., Verbruggen, M.J., Hill, T., Moffett, B., Fuerst, J., Schouten, S., Damsté, J.S.S., Harris, J., Shaw, P., Jetten, M., Strous, M., 2003. Candidatus "Scalindua brodae", sp. nov., Candidatus "Scalindua wagneri", sp. nov., two new species of anaerobic ammonium oxidizing bacteria. Syst. Appl. Microbiol. 26, 529–538. doi:10.1078/072320203770865837

Schmid, M.C., Risgaard-Petersen, N., Van De Vossenberg, J., Kuypers, M.M.M., Lavik, G., Petersen, J., Hulth, S., Thamdrup, B., Canfield, D., Dalsgaard, T., 2007. Anaerobic ammonium-oxidizing bacteria in marine environments: widespread occurrence but low diversity. Environ. Microbiol. 9, 1476–1484.

Schmidt, I., & Bock, E. (1997). Anaerobic ammonia oxidation with nitrogen dioxide by Nitrosomonas eutropha. *Archives of Microbiology*, *167*(2), 106–111. https://doi.org/10.1007/s002030050422

Sekiguchi, Y., Kamagata, Y., Syutsubo, K., Ohashi, a., Harada, H., & Nakamura, K. (1998). Phylogenetic diversity of mesophilic and thermophilic granular sludges determined by 16 rRNA gene analysis. *Microbiology*, *144*(1998), 2655–2665. https://doi.org/10.1099/00221287-144-9-2655

Sen, D., Mitta, P., & Randall, C. W. (1994). Performance of fixed film media integrated in activated sludge reactors to enhance nitrogen removal. *Water Science and Technology*, *30*(11), 13 LP-24. Retrieved from http://wst.iwaponline.com/content/30/11/13.abstract

Seo, Y.Y., 2009. Monitoring the role of biofilm biopolymers against disinfectants in water distribution systems. Water Resour. Cent. Annu. Tech. Rep. FY 2009.

Shannon, J.M., 2014. Partial nitritation-anammox using pH-controlled aeration in submerged attached growth bioreactors. The University of Iowa.

She, Z., Zhao, L., Zhang, X., Jin, C., Guo, L., Yang, S., Zhao, Y. and Gao, M., 2016. Partial nitrification and denitrification in a sequencing batch reactor treating high-salinity wastewater. Chemical Engineering Journal, 288, pp.207-215.

Shen, Q.-H., Jiang, J.-W., Chen, L.-P., Cheng, L.-H., Xu, X.-H., & Chen, H.-L. 2015. Effect of carbon source on biomass growth and nutrients removal of Scenedesmus obliquus for wastewater advanced treatment and lipid production. *Bioresource Technology*, *190*, 257–263.

Sheng, G.-P., Yu, H.-Q., Li, X.-Y., 2010. Extracellular polymeric substances (EPS) of microbial aggregates in biological wastewater treatment systems: A review. Biotechnol. Adv. 28, 882–894. doi:10.1016/j.biotechadv.2010.08.001

Siegrist, H., Salzgeber, D., Eugster, J., Joss, A., 2008. Anammox brings WWTP closer to energy autarky due to increased biogas production and reduced aeration energy for N-removal. Water Sci Technol. 57(3), 383-388.

Singh, N. K., & Kazmi, A. A. 2016. Environmental performance and microbial investigation of a single stage aerobic integrated fixed-film activated sludge (IFAS) reactor treating municipal wastewater. *Journal of Environmental Chemical Engineering*, 4(2), 2225–2237. https://doi.org/10.1016/j.jece.2016.04.001

Sonthiphand, P., Hall, M.W., Neufeld, J.D., 2014. Biogeography of anaerobic ammonia-oxidizing (anammox) bacteria. Front. Microbiol. 5. doi:10.3389/fmicb.2014.00399

Speth, D.R., Guerrero-Cruz, S., Dutilh, B.E., Jetten, M.S.M., 2016. Genome-based microbial ecology of anammox granules in a full-scale wastewater treatment system. Nat. Commun. 7.

Sriwiriyarat, T., Pittayakool, K., Fongsatitkul, P. and Chinwetkitvanich, S., 2008. Stability and capacity enhancements of activated sludge process by IFAS technology. Journal of Environmental Science and Health Part A, 43(11), pp.1318-1324.

Steuernagel, L., de Léon Gallegos, E. L., Azizan, A., Dampmann, A.-K., Azari, M., & Denecke, M. 2018. Availability of carbon sources on the ratio of nitrifying microbial biomass in an industrial activated sludge. *International Biodeterioration & Biodegradation*, *129*, 133–140.

Stijn, W., H., V. H. S. W., Pascal, B., P., V. E. I., Van, C. O., A., V. P., & Willy, V. 2004. Modeling and simulation of oxygen-limited partial nitritation in a membrane-assisted bioreactor (MBR). *Biotechnology and Bioengineering*, *86*(5), 531–542. https://doi.org/doi:10.1002/bit.20008

Strous, M., & Jetten, M. 1997. Effects of aerobic and microaerobic conditions on anaerobic ammonium-oxidizing ( anammox ) sludge . Effects of Aerobic and Microaerobic Conditions on Anaerobic Ammonium-Oxidizing ( Anammox ) Sludge. *Applied and Environmental Microbiology*, *63*(6), 2446–2448.

Strous, M., Heijnen, J. J., Kuenen, J. G., & Jetten, M. S. M. 1998. The sequencing batch reactor as a powerful tool for the study of slowly growing anaerobic ammonium-oxidizing microorganisms. *Applied Microbiology and Biotechnology*, *50*(5), 589–596. https://doi.org/10.100 //s002530051340

Strous, M., Kuenen, J. G., & Jetten, M. S. M. 1999. Key Physiology of Anaerobic Ammonium Oxidation Key Physiology of Anaerobic Ammonium Oxidation. *Applied and Environmental Microbiology*, *65*(7), 0–3. https://doi.org/papers2://publication/uuid/E9A1573A-6D62-420E-94D0-CA7C84D0FEB9

Strous, M., Kuenen, J.G., Jetten, M.S.M., 1999. Key physiology of anaerobic ammonium oxidation. Appl. Environ. Microbiol. 65, 3248–3250.

doi:papers2://publication/uuid/E9A1573A-6D62-420E-94D0-CA7C84D0FEB9

Suneethi, S., Sri Shalini, S., Joseph, K., 2014. State of The Art Strategies for Successful ANAMMOX Startup and Development: A Review. Int J Waste Resour. 4, 2. doi:10.4172/2252-5211.1000168

Szatkowska, B., Cema, G., Plaza, E., Trela, J., & Hultman, B. 2007. A one-stage system with partial nitritation and Anammox processes in the moving-bed biofilm reactor. *Water Science and Technology, 55*(8–9), 19 LP-26. Retrieved from http://wst.iwaponline.com/content/55/8-9/19.abstract

Takekawa, M., Park, G., Soda, S., Ike, M., 2014. Simultaneous anammox and denitrification (SAD) process in sequencing batch reactors. Bioresource Technol. 174, 159-166.

Talebizadeh, M., Belia, E., Vanrolleghem, P.A., 2014. Probability-based design of wastewater treatment plants. In Proceedings: International Congress on Environmental Modeling and Software (IEMSs2014). San Diego, California, USA,

Tam, N.F.Y., Wong, Y.S., Leung, G., 1992. Effect of exogenous carbon sources on removal of inorganic nutrient by the nitrification-denitrification process. Water Res. 26, 1229–1236. doi:10.1016/0043-1354(92)90183-5

Tan, C.H., Lee, K.W.K., Burmølle, M., Kjelleberg, S. and Rice, S.A., 2017. All together now: experimental multispecies biofilm model systems. Environmental microbiology, 19(1), pp.42-53.

Tsushima, I., Ogasawara, Y., Kindaichi, T., Satoh, H., Okabe, S., 2007. Development of high-rate anaerobic ammonium-oxidizing (anammox) biofilm reactors. Water Res, 41(8), 1623-1634.

Tchobanoglous, G., Burton, F. L., Stensel, H. D., & Eddy, M. &. 2003. *Wastewater Engineering: Treatment and Reuse*. McGraw-Hill. Retrieved from https://books.google.de/books?id=_iBSAAAAMAAJ

Thamdrup, B., & Dalsgaard, T. 2002. Production of N 2 through Anaerobic Ammonium Oxidation Coupled to Nitrate Reduction in Marine Sediments Production of N 2 through Anaerobic Ammonium Oxidation Coupled to Nitrate Reduction in Marine Sediments. *Applied and Environmental Microbiology, 68*(3), 1312–1318. https://doi.org/10.1128/AEM.68.3.1312

Third, K.A., Sliekers, A.O., Kuenen, J.G., Jetten, M.S.M., 2001. The CANON system (completely autotrophic nitrogen-removal over nitrite) under ammonium limitation: interaction and competition between three groups of bacteria. Syst. Appl. Microbiol. 24,

588–596.

Tomaszewski, M., Cema, G., & Ziembińska-Buczyńska, A. 2017. Influence of temperature and pH on the anammox process: A review and meta-analysis. *Chemosphere*, *182*, 203–214. https://doi.org/10.1016/j.chemosphere.2017.05.003

Torà, J. A., Lafuente, J., Baeza, J. A., & Carrera, J. 2010. Combined effect of inorganic carbon limitation and inhibition by free ammonia and free nitrous acid on ammonia oxidizing bacteria. *Bioresource Technology*, *101*(15), 6051–6058.

Tran, H. T., Park, Y. J., Cho, M. K., Kim, D. J., & Ahn, D. H. 2006. Anaerobic ammonium oxidation process in an upflow anaerobic sludge blanket reactor with granular sludge selected from an anaerobic digestor. *Biotechnology and Bioprocess Engineering*, *11*(3), 199–204. https://doi.org/Doi 10.1007/Bf02932030

Tsilogeorgis, J., Zouboulis, A., Samaras, P., & Zamboulis, D. 2008. Application of a membrane sequencing batch reactor for landfill leachate treatment. *Desalination*, *221*(1–3), 483–493. https://doi.org/10.1016/j.desal.2007.01.109

Tsushima, I., Ogasawara, Y., Kindaichi, T., Satoh, H., Okabe, S., 2007. Development of high-rate anaerobic ammonium-oxidizing (anammox) biofilm reactors. Water Res. 41, 1623–1634. doi:10.1016/j.watres.2007.01.050

Turk, O., & Mavinic, D. S. 1989. Maintaining nitrite build-up in a system acclimated to free ammonia. *Water Research*, *23*(11), 1383–1388. https://doi.org/10.1016/0043-1354(89)90077-8

Tyagi, R.D. and Vembu, K., 1990. Wastewater treatment by immobilized cells. CRC press.

U.S.Environmental Protection Agency. (2015). Case Studies on Implementing Low-Cost Modifications to Improve Nutrient Reduction at Wastewater Treatment Plants. https://www.epa.gov/nutrient-policy-data/case-studies-implementing-low-cost-modifications-improve-nutrient-reduction

Van de Graaf, A.A., Mulder, A., De Bruijn, P., Jetten, M.S.M., Robertson, L.A., Kuenen, J.G., 1995. Anaerobic oxidation of ammonium is a biologically mediated process. Appl. Environ. Microbiol. 61, 1246–1251. doi:PMC167380

Van der Star, W.R.L., Abma, W.R., Blommers, D., Mulder, J.-W., Tokutomi, T., Strous, M., Picioreanu, C., van Loosdrecht, M.C.M., 2007. Startup of reactors for anoxic ammonium oxidation: experiences from the first full-scale anammox reactor in Rotterdam. Water Res. 41, 4149–4163.

Van Dongen, U.G.J.M., Jetten, M.S., Van Loosdrecht, M.C.M., 2001. The SHARON®-Anammox® process for treatment of ammonium rich wastewater. Water Sci Technol,

44(1), 153-160.

Van Hulle, S.W., Vandeweyer, H.J., Meesschaert, B.D., Vanrolleghem, P.A., Dejans, P., Dumoulin, A., 2010. Engineering aspects and practical application of autotrophic nitrogen removal from nitrogen rich streams. Chem Eng J. 162(1), 1-20.

Van Hulle, S. W. H., Volcke, E. I. P., Teruel, J. L., Donckels, B., van Loosdrecht, M., & Vanrolleghem, P. A. 2007. Influence of temperature and pH on the kinetics of the Sharon nitritation process. *Journal of Chemical Technology and Biotechnology*, *82*(5), 471–480.

van Kessel, M.A.H.J., Speth, D.R., Albertsen, M., Nielsen, P.H., Op den Camp, H.J.M., Kartal, B., Jetten, M.S.M., Lücker, S., 2015. Complete nitrification by a single microorganism. Nature 528, 555–559. doi:10.1038/nature16459

van Loosdrecht, M. C. M., Eikelboom, D., Gjaltema, A., Mulder, A., Tijhuis, L., & Heijnen, J. J. 1995. Biofilm structures. *Water Science and Technology*, *32*(8), 35 LP-43. Retrieved from http://wst.iwaponline.com/content/32/8/35.abstract

Van Loosdrecht, M. C. M., Heijnen, J. J., Eberl, H., Kreft, J., & Picioreanu, C. 2002. Mathematical modeling of biofilm structures. *Antonie van Leeuwenhoek, International Journal of General and Molecular Microbiology*, *81*(1–4), 245–256. https://doi.org/10.1023/A:1020527020464

van Loosdrecht, M.C. and Brdjanovic, D., 2014. Anticipating the next century of wastewater treatment. Science, 344(6191), pp.1452-1453.

van Loosdrecht, M.C.M., Lopez-Vazquez, C.M., Meijer, S.C.F., Hooijmans, C.M., Brdjanovic, D., 2015. Twenty-five years of ASM1: past, present and future of wastewater treatment modeling. J. Hydroinformatics 17, 697–718. doi:10.2166/hydro.2015.006

van Niftrik, L.A., Fuerst, J.A., Damsté, J.S.S., Kuenen, J.G., Jetten, M.S., Strous, M., 2004. The anammoxosome: an intracytoplasmic compartment in anammox bacteria. FEMS Microbiol Lett. 233(1), 7-13.

Van Niftrik, L.A., Jetten, M.S.M., 2012. Anaerobic Ammonium-Oxidizing Bacteria: Unique Microorganisms with Exceptional Properties. Microbiol. Mol. Biol. Rev. 76, 585–596. doi:10.1128/MMBR.05025-11

vandeGraaf, A.A., deBruijn, P., Robertson, L.A., Jetten, M.S.M., Kuenen, J.G., 1997. Metabolic pathway of anaerobic ammonium oxidation on the basis of N-15 studies in a fluidized bed reactor. Microbiology-Uk 143, 2415–2421. doi:10.1099/00221287-143-7-2415

Verstraete, W., Wittebolle, L., Heylen, K., Vanparys, B., de Vos, P., van de Wiele, T., &

Boon, N. (2007). Microbial Resource Management: The road to go for environmental biotechnology. *Engineering in Life Sciences*, *7*(2), 117–126. https://doi.org/10.1002/elsc.200620176

Villaverde, S., García-Encina, P. A., & Fdz-Polanco, F. 1997. Influence of pH over nitrifying biofilm activity in submerged biofilters. *Water Research*, *31*(5), 1180–1186. https://doi.org/10.1016/S0043-1354(96)00376-4

Vlaeminck, S. E., Cloetens, L. F. F., Carballa, M., Boon, N., & Verstraete, W. 2008. Granular biomass capable of partial nitritation and anammox. *Water Science and Technology*, *58*(5), 1113 LP-1120. Retrieved from http://wst.iwaponline.com/content/58/5/1113.abstract

Vlaeminck, S.E., De Clippeleir, H., Verstraete, W., 2012. Microbial resource management of one-stage partial nitritation/anammox. Microb. Biotechnol. 5, 433–448.

Wang, J.. Gu, J.D., 2013. Dominance of Ca. Scalindua species in anammox community revealed in soils with different duration of rice paddy cultivation in Northeast China. Appl Microbiol Biot. 97(4), 1785-1798

Wang, Z.W., Liu, Y., Tay, J.H., 2005. Distribution of EPS and cell surface hydrophobicity in aerobic granules. Appl. Microbiol. Biotechnol. 69, 469–473. doi:10.1007/s00253-005-1991-5

Wanner, O. and Morgenroth, E., 2004. Biofilm modeling with AQUASIM. Water Science and Technology, 49(11-12), pp.137-144.

Wanner, O. and Reichert, P., 1996. Mathematical modeling of mixed-culture biofilms. Biotechnology and bioengineering, 49(2), pp.172-184.

Weissbrodt, D.G., Neu, T.R., Kuhlicke, U., Rappaz, Y., Holliger, C., 2016. Assessment of bacterial and structural dynamics in aerobic granular biofilms. Bioremediation of Wastewater: Factors and Treatment. p141.

Wenjie, Z., Huaqin, W., Joseph, D.R., Yue, J., 2015. Granular activated carbon as nucleus for formation of anammox granules in an expanded granular-sludge-bed reactor. Global Nest J, 17(3), 508-514.

Wingender, J., Neu, T.R., Flemming, H.-C., 1999. What are bacterial extracellular polymeric substances?, in: Microbial Extracellular Polymeric Substances. pp. 1–19. doi:10.1007/978-3-642-60147-7_1

Winkler, M.-K., Kleerebezem, R., & Van Loosdrecht, M. C. M. 2012. Integration of anammox into the aerobic granular sludge process for main stream wastewater treatment at ambient temperatures. *Water Research*, *46*(1), 136–144.

Xie, W.M., Ni, B.J., Seviour, T., Sheng, G.P., Yu, H.Q., 2012. Characterization of autotrophic and heterotrophic soluble microbial product (SMP) fractions from activated sludge. Water Res. 46, 6210–6217. doi:10.1016/j.watres.2012.02.046

Yang, J., Trela, J., Zubrowska-sudol, M., & Plaza, E. 2015. Intermittent aeration in one-stage partial nitritation / anammox process. *Ecological Engineering*, *75*, 413–420. https://doi.org/10.1016/j.ecoleng.2014.11.016

Yang, Q., Wang, S., Yang, A., Guo, J., & Bo, F. 2007. Advanced nitrogen removal using pilot-scale SBR with intelligent control system built on three layer network. *Frontiers of Environmental Science & Engineering in China*, *1*(1), 33–38.

Ye, J., McDowell, C. S., Koch, K., Kulick, F. M., & Rothermel, B. C. 2009. Pilot Testing of Structured Sheet Media IFAS for Wastewater Biological Nutrient Removal (BNR). *Proceedings of the Water Environment Federation*, *2009*(12), 4427–4442.

Yokota, N., Watanabe, Y., Tokutomi, T., Kiyokawa, T., Hori, T., Ikeda, D., Song, K., Hosomi, M. and Terada, A., 2017. High-rate nitrogen removal from waste brine by marine anammox bacteria in a pilot-scale UASB reactor. Applied microbiology and biotechnology, pp.1-12.

Zarda, B., Amann, R., Wallner, G., Schleifer, K.-H., 1991. Identification of single bacterial cells using digoxigenin-labelled, rRNA-targeted oligonucleotides. Microbiology 137, 2823–2830.

Zhang, L., Liu, M., Zhang, S., Yang, Y., Peng, Y., 2015. Integrated fixed-biofilm activated sludge reactor as a powerful tool to enrich anammox biofilm and granular sludge. Chemosphere. 140, 114-118.

Zhang, L., Narita, Y., Gao, L., Ali, M., Oshiki, M., Okabe, S., 2017. Maximum specific growth rate of anammox bacteria revisited. Water Res. 116, 296–303.

Zhu, A., Guo, J., Ni, B.-J., Wang, S., Yang, Q., Peng, Y., 2016. A novel protocol for model calibration in biological wastewater treatment, in: Environmental Engineering and Activated Sludge Processes: Models, Methodologies, and Applications. Apple Academic Press, pp. 23–47.

Zhu, G., Peng, Y., Li, B., Guo, J., Yang, Q., & Wang, S. (2008). Biological removal of nitrogen from wastewater. *Rev Environ Contam Toxicol*, *192*(January 2014), 159–195.

# Appendices

## Appendix A.

### Optimization of quantitative fluorescence in situ hybridization (qFISH)

Using fluorescence in situ hybridization (qFISH) method, which fluorescent-labeled oligonucleotide probes to 16S rRNA combines with digital image analysis to measure the microbial abundance within the biofilm. The protocol focused on steps bellow that follow the FISH procedure and lead to quantitative results.

*Samples collection and pre-treatment*

Five granular biofilm samples were collected from ZDE plant in monthly basis from end-July to mid-December in 2016. In order to optimize the qFISH protocol, the different mechanical and chemical pre-treatment methods were proposed and tested on the first batch of granular samples.

Preparation for mechanical pre-treatment by glass beads (marked as number 1) and ultrasonic (marked as number 2) was done (Fig. 11). For each step 5 ml of granular sample was filled in a 50 ml Falcon™ conical centrifuge tube (VWR International, USA), and mixed with 10 ml of distill water. During the pre-treatment method using glass beads (1), 10 ml glass pearl beads with $5 \pm 0.3$ mm diameter (Carl Roth GmbH + Co. KG, Germany) is added into one tube, and the suspension was vortexed for 2 minutes. During the ultrasonic pretreatment (2), another tube was under ultrasonic waveswith a sonicator bath (Schollsonic 4000, EMAG AG, Germany) for 5 minutes.

After that, both mechanically treated samples (1 and 2) were (A) non-washed, (B) washed with distilled water or (C) chemically treated with cyclohexane (Carl Roth GmbH + Co. KG, Germany) (Fig.11). With regards to washing with water, 2 ml sample (1B & 2B) was filled in 2ml microcentrifuge tube (VWR Internationals, USA), and centrifuged by a laboratory centrifuge at the speed of 14000 round per minute (rpm), 4 °C for 15 minutes (Centrifuge 5810R, Eppendorf, Germany). The supernatant afterward was discarded, following 2 ml of distilled water was added, and then the pellet was mixed into the water. The centrifugation steps were repeated 3 times.

With regards to samples treated chemically with cyclohexane (1C & 2C), 10 ml of sample, 20 ml of cyclohexane and 20 ml of distilled water were completely mixed in a 100 ml SCHOTT DURAN® laboratory bottle (VWR Internationals, USA). The cyclohexane and aqueous phases were separated using a separator funnel. After 15-minute interval in the funnel, 2 layers were split from each other, cyclohexane and cells were the upper layer and the liquid water was the lower layer. First, aqueous phase was extracted and 2 ml of water sample (1C, 2C water extraction) was inoculated in the 2ml microcentrifuge tube for further cell fixation. The cyclohexane phase was mixed with 20 ml distilled water and the extraction procedure was repeated for 3 times. At the end, the final emulsion of cyclohexane and extracted cells was centrifuged at 14000 rpm, 4 °C in 15 minutes to remove the cyclohexane. The pellets were washed in 15 ml distilled water and centrifuged at 14000 rpm, 4 °C in 10 minutes. After discarding the supernatant, the final pellets were mixed in 2 ml distilled water (1C, 2C cyclohexane extraction) in the 2ml microcentrifuge tube for further cell fixation. All pre-treated samples (including the non-washed samples 1A and 2A) will be analyzed. The results will compare and detect the best pre-treatment method for qFISH protocol.

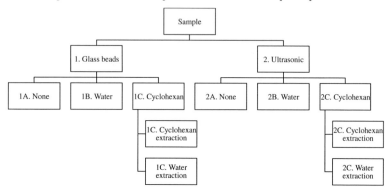

Figure A1: Scheme of sample pre-treatment methods for optimization of qFISH

*Establishment of optimized qFISH protocol*

In order to optimize the qFISH protocol, different pre-treatment methods including mechanical pretreatment by glass beads or ultrasonic and chemical treatment by cyclohexane were applied. In total, six different alternative pre-treatment methods were evaluated (Table. 6), for each method five independent oligonucloide specific FISH probe and one general domain probe (EUB338) were applied. Multiple numbers of possible randomized FOVs (minimum 20) were taken for each specific probe and then the congruency factor, which is the overlapping level between signal from specific probe and general probe was analyzed by daime 2.1 sofware. The congruency of each method was used as criteria to choose the best pre-treatment method for qFISH protocol, the weighted average of congruency factor was calculated in Fig. 15.

Table A1: Different pre-treatment methods for optimization qFISH protocol

| Pre-treatment method | None-washed | Washed with water | Pre-treatment with cyclohexane |
|---|---|---|---|
| Mechanical pre-treatment by glass beads | 1 A | 1 B | 1 C |
| Mechanical pre-treatment by ultrasonic | 2 A | 2 B | 2 C |

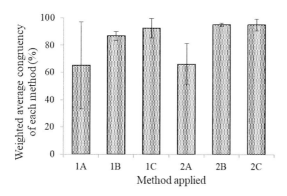

Figure A2: Weighted average of congruency factor for different methods

The results indicated that methods without any washing steps (1A and 2A) had lowest congruent percentages, while other methods could reach more than 85 % of congruency. Application of glass beads or ultrasonic on granular samples was to disaggregate cells and re-distributed homogenously cells into each sample. In order to remove any cell lysis or excess inorganic substances washing step was important. Average congruency of samples (1B and 2B) was washed with water after mechanical pre-treatment possessed 86.8 ± 3.3 % and 95.1 ± 1.3 % respectively, which was approximately 20 % higher than samples (1A and 2A) without washing (Fig.15). The second pre-treatment method was proposed to eliminate excess amount of water in the samples, which applied extraction method. The cyclohexane solvent was chosen since cyclohexane is immiscible in water, with lower density (0.7781 g mL$^{-1}$) than water (1.0 g mL$^{-1}$). Results showed that there was very good overlap between red signal for specific probes and green signal for general probe. The sample pre-treated with glass bead and extracted by cyclohexane (1C) indicated 92.6 ± 7.0 % of congruency, and the congruency result of the sample treated by ultrasonic and extracted by cyclohexane (2C) was 95.0 ± 4.3 %. (Fig.15).

The study from Almstrand et al (2014) introduced minimum congruency level of image processing for qFISH was at least 90 % (Almstrand et al., 2014). Thus, comparing results of weighted average congruency, three methods 1C, 2B and 2C had the highest agreement. The pre-treatment method with ultrasonic and wash with water – 2B was chosen to be final pre-treatment method for qFISH protocol (table 5A, appendix) to verify the bio-volume fraction of granular samples due to following reasons.

The application of glass beads to remove and homogenize thick biofilm layers usually is combined with other mechanical force such as scraping and followed by ultrasonic, but using only glass beads may not feasible in many other cases (Heitz et al., 1997). In contrast, ultrasonic treatment has been widely used and achieved the best results of removal and survival of cells. It should be taken into account that sonication time should be from 1 to 5 minutes with frequency of 40 – 50 kHz (Flemming et al., 2000). Finally, though 2B and 2C methods indicated similar results, pre-treatment method 2B required less complicated procedure, less time consuming, and possessed the lowest standard deviation. Hence, the mechanical pre-treatment by ultrasonic followed by washing with water was used for all qFISH samples in this study.

*Cell fixation and washing procedure*

Each 500 µl of samples was mixed with 1500 µl of Roti®-Histofix 4 % containing formaldehyde solution 4% with the pH 7(Carl Roth GmbH + Co. KG, Germany) with the dilution factor 1:3 in a 2 ml microcentrifuge tube. The fixation duration could last from 2 to 24 hours. In the next step, samples were washed with water. 2 ml of fixed cells in formaldehyde solution was centrifuged at 10000 rpm, 4 °C in 10 minutes, then 1500 µl of supernatant was discarded and 1500 µl of phosphate buffered saline (PBS) 10 % (Thermo Fisher Scientific, USA) was added into each tube for rinsing. These steps were repeated 3 times. During the last washing step, 750 µl of supernatant was removed instead of 1500 µl, then a mixture of 250 µl PBS 10 % and 500 µl ethanol 96 % (Carl Roth GmbH + Co. KG, Germany) is added into each sample. The detailed overview on the fixation and washing steps are presented in Table 5A in the appendix.

| | | |
|---|---|---|
| 1. Pre-treatment | | • Add 5 mL granules in to a PE tube 15 mL<br>• Mix and vortex with 5 mL water<br>• Place the sample in ultrasonic bath for 5 mins |
| 2. Fixation | | • Fill the sample in 2.0 mL microcentrifuge Eppendorf tubes |
| | | • Centrifuge at 4000 rpm, 4°C in 15 mins. Afterward remove supernatant, and add distilled water<br>• Mix the solution, and repeat above step 3 times |
| | Histofix 4 %<br>Suspended cells | • After washing with water, 500 μL suspended cells are mixed with 1500 μL Roti-Histofix 4 %. (or any other standard fixative solution with Paraformaldehyde i.e. PFA 4%.).<br>• 2 to 3 h incubation is required. (The incubation time can be increased up to maximum overnight in certain cases)<br>• Fixed sample is stored in the fridge over night |
| 3. Washing procedure | • 2ml fixed samples in Eppendorf tubes are centrifuged at 1000 rpm 4°C in 10 min.<br>• Discard 1500 μL supernatant<br>• Add 1500 μL PBS 10 %<br>Repeat above steps for three times.<br>After the he last washing step, discard 750 μL supernatant<br>• Add 250 μL PBS 10 % and 500 μL ethanol 96 %<br>• Re-suspend and store washed samples in the fridge (better to be stored in the freezer to be usable up to several months) | |

| 4. Slide preparation | • Slides must be cleaned with ethanol<br><br>Record the sample distribution by pencil in the slides and note details (date, sample name, gene probe name...) in the lab diary |
|---|---|
| 5. Hybridization buffer | Choose hybridization buffers corresponding to the formamide concentration (see in the FISH protocol). The formamide concentration corresponding to the gene probe can be found in www.probebase.net<br><br>Note: - The hybridization buffers are prepared in 50 mL PE tubes<br><br>        - SDS 10 % is added as the last step |
| 6. Application of sample | • Add 10 mL suspended cells in each well, dry in oven at 46°C<br><br>Repeat this step two to three time to increase higher amount of biomass on the slide<br><br>• Dehydrate cells by dipping glass slides in ethanol solutions following order of 50 %, 80 % and 96 % for 5 minutes (see details in standard FISH protocol)<br><br><br><br>The first application of biomass on a glass slide     Increase the application of biomass on the glass slide to prevent cells loss on the glass surface<br><br>• Dry the slides in the oven at 46°C |
| 7. Application of gene probe | • The domain probe (EUB 338 or mix-EUB338) and specific target probe are mixed with Te-buffer and aliquotes of 10 μL in micro Eppendorf tubes are stored in the freezer (see aliquoting protocol)<br><br>• Add 40 ml of hybridization buffer to the domain probe and target probes separately. The final volume of each probe reaches 50 μL<br><br>• Mix the solutions<br><br>• Apply 5 μL of the domain probe, following by 5 μL specific target probe<br><br>• Use the tip of pipet to mix the solutions (do not touch the surface of slides!)<br><br>• In case of applying single gene probe, see FISH protocol |
| 8. Hybridization | • The hybridization buffer tubes are place horizontally in the metal rack (maximum 8 tubes per rack)<br><br>• Insert a piece of tissue then the slide on top of the tissue, facing upwards<br><br>• The water bath is pre-heated at 46°C (check the actual temperature of the water bath by using thermometer)<br><br>• Place the metal rack horizontally in the water bath<br><br>• Incubate over night (from 20 to 24 h) (see the FISH protocol) |
| 9. Washing buffer | • Washing buffers are prepared base on the formamide concentration (see FISH protocol and www.probebase.net) |

| | |
|---|---|
| | • The washing buffers can prepared together with hybridization buffers and pre-heated by incubate in water bath at 48°C |
| 10. Counterstaining DAPI (optional) | See details in FISH protocol Note: This step can be useful to check the hybridization efficiency but it is optional. |
| 11. Application of cover slip | See details in FISH protocol |
| 12. Light microscopy and image acquisition | • Images are captured at low magnification (100x or 400x) in order to capture more biomass in one picture.<br>• Image acquisition is done by software AxioVision 4.18<br>• Field of views (FOVs) must be recorded randomly at chosen position within the sample<br>• Make sure that the target and general cells are congruent in each image pair<br>• Save the image pair in TIFF file and in order (e.g. image000, image 001…) |
| 13. Image processing | <br>• Download and open image J software, which is an open source, available on https://imagej.net/Downloads<br>• Open file by **File -> Import -> TIFF virtual stack**<br>• The FISH images appear in the form of "image pair" in different colors<br>• Split the pair image (into red, green and blue) by click to **Image -> Color -> Split channels**. The software will show 8 bit greyscale images in single signal (red, green and blue). In the example, the images contains green and red signals. |

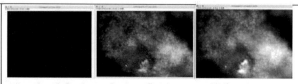

- To reduce the background noise of the images, use the function:

**Process -> subtract background**

Depending on images that the rolling ball radius pixels can vary from 15 to 250.

(If the radius pixels are too low, picture can underestimated and vice versa.)

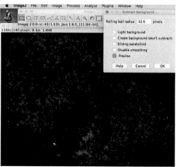

- Set a threshold manually for each image, active threshold by selecting on

**Image -> adjust -> threshold**

The histogram represents the distribution of pixels intensity in an image.

A pixel with an intensity of 0 is black and a value of 225 is white.

- Drag the slider to adjust the intensity of the image in a way that the cells are in shade of grey and background is black.

- Save the threshold function by clicking on **Apply**.

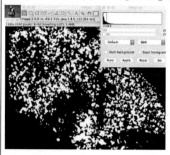

- Save the images **File -> Save as -> Tiff...**

| | Note: the images in a pair must be saved in the same order (e.g. green-image000 and red-image000) for further image analysis |
|---|---|
| 14. Image analysis | • Use DAIME 2.1 software<br><br>In order to get reliable results, each probe target well should take randomly from 20 to 30 fields of view (FOV). Each FOV in form of "pair images" are set the threshold manually and saved in greyscale images and in the TIFF image format (8 bits per pixel)<br><br>• Load the greyscale image series of the green and red signal **File -> import**. All series must be saved respectively (from 000 till end), so that the software is able to open all images.<br><br>• Segment images automatically by selecting AUTO in the main menu. Active options like the picture below, and click **Segment!**<br><br>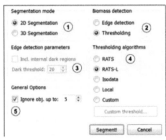<br><br>• The artifact rejection can be done by clicking to the segment picture series and choosing **Object -> Select artifact -> choose Reject from Object tools -> Save -> Ok**<br><br>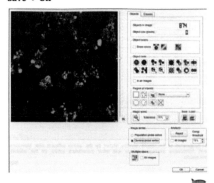<br><br>• For estimation of biovolume fraction, click to VOL% in the main toolbar, report to the dialog which series is the general signal and which one is the target |

probe -> **Ok.** The software will measure biovolume fraction by device the pixels belonging to target image over general image.

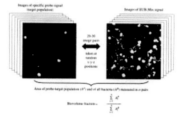

- The result includes i) average biovolume fraction ii) pixel sums of each signals (green, red or blue) and iii) average congruency (must be more than 90 %). An example of biovolume fraction is shown in the picture below, and allows exporting the data to excel file.

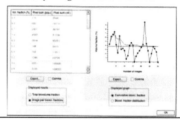

Figure A3: FISH and qFISH protocol

**Appendix B**


Table B1: The comparison of NRE to show how the plant has been improved after two process in activated sludge tank and activated carbon beds were combined.

| NRE (%) | | 2007 | 2008 | 2009 | 2010 | 2011 | 2012 | 2013 | 2014 | 2015 |
|---|---|---|---|---|---|---|---|---|---|---|
| Final NRE (%) Combined [*] | Mean | 93.3 | 92.5 | 90.7 | 93.0 | 93.9 | 93.5 | 94.2 | 94.3 | 92.8 |
| Biological NRE (%) Only activated sludge | Mean | 76.3 | 88.3 | 69.5 | 86.2 | 85.1 | 84.4 | 86.3 | 86.0 | 84.4 |


*The Table shows a comparison of removal efficiency to show how the plant has been optimized and how much is the effect of two combined process in activated sludge tank and activated carbon beds. The best average efficiency on daily data was reached during 2014 when the NRE for activated sludge system was 86 % and for final outflow was averagely reached to 94.3 % which is the highest reported NRE of the plant. The weakest efficiencies were observed during 2009 however during this period the highest compensation of nutrient was achieved after activated carbon process. By considering the treatment by granules in the activated carbon, NE increased from 69.5 % to 89.7% which reflects a raise of NRE by 22.2 %.

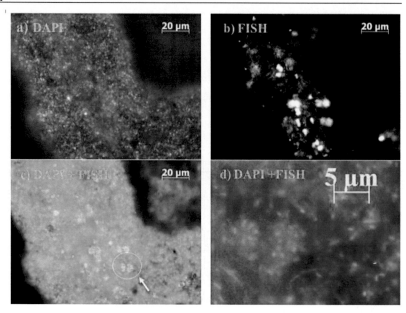

Fig. B1: FISH micrographs of an floccular biomass sample fixed with 4% PFA as fixative and hybridized with 5% formamide FISH buffer containing *Candidatus* Scalindua using specific probe to target genus *Candidatus* Scalindua a type of anammox bacteria i.e. S-G-Sca-1309-a-A-21. Planctomycetes including Candidatus "Scalindua wagneri" are labeled pink and DAPI is showed with turquoise-blue color.

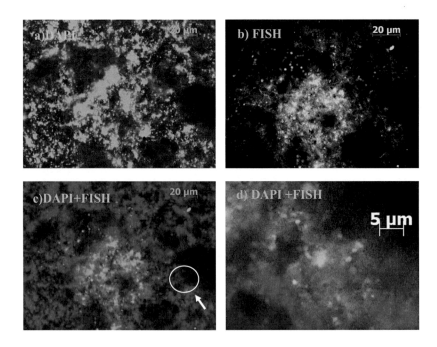

Fig. B2: FISH micrographs of an biomass sample fixed with 4% PFA as fixative and hybridized with 15% formamide FISH buffer containing *Candidatus* Brocadiaceae using specific probe to target all ANAMMOX bacteria i.e. S-*-Amx-0368-a-A-18 (a,e) DAPI staining under light microscopy converted to grayscale (b,f) FISH staining results under SEM converted to grayscale (c,d,g and h) Signals from specific probe labelled with 6FAM (green) is overlapping with DNA contained cells stained by DAPI with blue signals result in *Candidatus* Brocadiaceae that appear turquoise (c,g) represents for lower magnification showing all stained areas but in the image (d,h) the marked layers shown by white arrow in images (c,g) have been magnified for better representation of the shape and morphology of targeted organisms.

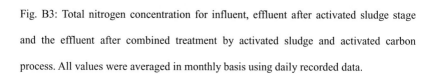

Fig. B3: Total nitrogen concentration for influent, effluent after activated sludge stage and the effluent after combined treatment by activated sludge and activated carbon process. All values were averaged in monthly basis using daily recorded data.

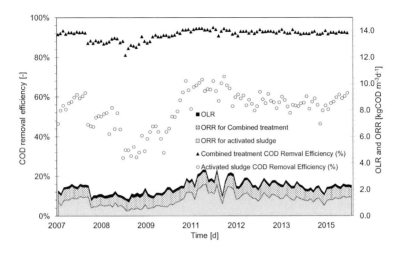

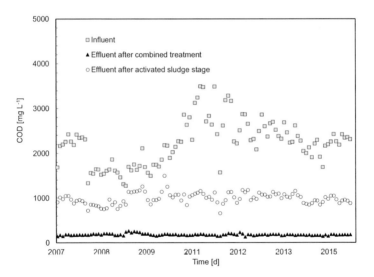

Fig. B4: NRE for biological and final treatment efficiency, NLR and NRR for only biological treatment with activated sludge and combined biological treatment by activated sludge with activated carbon process. All values are daily. Top image is for 2010 and bottom for year 2014. (Examples)

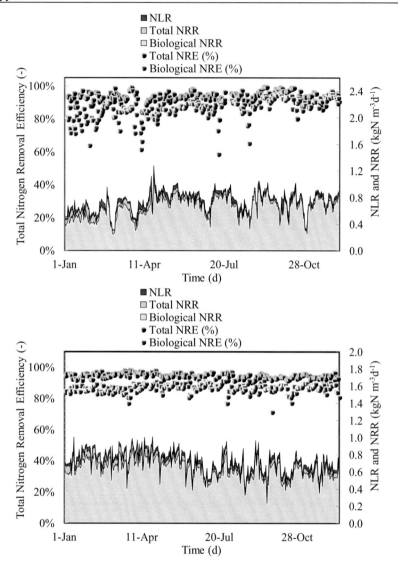

Fig. B5: Total nitrogen concentration for influent, effluent biological treatment with activated sludge and final effluent after combined biological treatment by activated sludge and activated carbon process. All valued were daily. Top image is for 2007 and bottom for year 2015. (Examples)

Appendix C

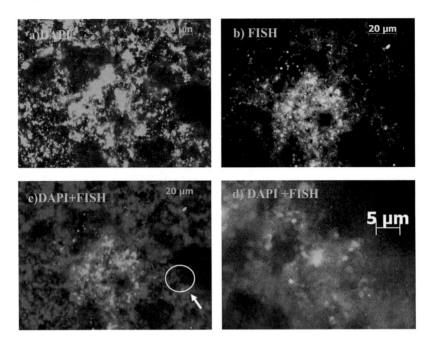

Fig. C1: FISH micrographs of an biomass sample fixed with 4% PFA as fixative and hybridized with 15% formamide FISH buffer containing *Candidatus* Brocadiaceae using specific probe to target all ANAMMOX bacteria i.e. S-*-Amx-0368-a-A-18 (a,e) DAPI staining under light microscopy converted to grayscale (b,f) FISH staining results under SEM converted to grayscale (c,d,g and h) Signals from specific probe labelled with 6FAM (green) is overlapping with DNA contained cells stained by DAPI with blue signals result in *Candidatus* Brocadiaceae that appear turquoise (c,g) represents for lower magnification showing all stained areas but in the image (d,h) the marked layers shown by white arrow in images (c,g) have been magnified for better representation of the shape and morphology of targeted organisms.

Table C1: Revised process kinetic rate equations using two group of major anammox species with distinct parameters i.e. *Ca.* Brocadia anammoxidans and *Ca.* Scalindua sp.

| Process | Kinetics rates equation |
| --- | --- |
| 1. Hydrolysis | $k_H \dfrac{X_S/X_H}{K_X + X_S/X_H} X_H$ |
| 2. Growth of anammox genus *Ca.* Scalindua | $\mu_{AN}^{Sca} \dfrac{K_{O2}^{AN}}{K_{O2}^{AN,Sca} + S_{O2}} \dfrac{S_{NH4}}{K_{NH4}^{AN,Sca} + S_{NH4}} \dfrac{S_{NO2}}{K_{NO2}^{AN,Sca} + S_{NO2}} X_{AN,Sca}$ |
| 3. Decay of anammox genus *Ca.* Scalindua | $b_{AN}^{Sca} X_{AN,Sca}$ |
| 4. Growth of anammox genus *Ca.* Brocadia | $\mu_{AN}^{Bro} \dfrac{K_{O2}^{AN}}{K_{O2}^{AN,Bro} + S_{O2}} \dfrac{S_{NH4}}{K_{NH4}^{AN,Bro} + S_{NH4}} \dfrac{S_{NO2}}{K_{NO2}^{AN,Bro} + S_{NO2}} X_{AN,Bro}$ |
| 5. Decay of anammox genus *Ca.* Brocadia | $b_{AN}^{Bro} X_{AN,Bro}$ |
| 6. Anoxic growth of $X_H$ using nitrite | $\mu_H \eta_{NOx} \dfrac{K_{O2}^{H}}{K_{O2}^{H} + S_{O2}} \dfrac{S_{NH4}}{K_{NH4}^{H} + S_{NH4}} \dfrac{S_{NO2}}{K_{NO2}^{H} + S_{NO2}} \dfrac{S_{NO2}}{S_{NO3} + S_{NO2}} \dfrac{S_S}{K_S + S_S} X_H$ |
| 7. Anoxic growth of $X_H$ using nitrate | $\mu_H \eta_{NO} \dfrac{K_{O2}^{H}}{K_{O2}^{H} + S_{O2}} \dfrac{S_{NH4}}{K_{NH4}^{H} + S_{NH4}} \dfrac{S_{NO3}}{K_{NO3}^{H} + S_{NO3}} \dfrac{S_{NO3}}{S_{NO2} + S_{NO3}} \dfrac{S_S}{K_S + S_S} X_H$ |
| 8. Decay of $X_H$ | $b_H X_H$ |

Table C2: Revised stoichiometric matrix for SAD model

| Process | $S_S$ | $S_{NH4}$ | $S_{NO2}$ | $S_{NO3}$ | $S_{N2}$ | $X_S$ | $X_{AN,Sca}$ | $X_{AN,Bro}$ | $X_H$ | $X_I$ |
|---|---|---|---|---|---|---|---|---|---|---|
| | COD | N | N | N | N | COD | COD | COD | COD | COD |
| 1 | 1 | | | | | -1 | | | | |
| 2 | | $-i_{NBM}-1/Y_{AN}^{Sca}$ | $-0.87/Y_{AN}^{Sca}-0.76$ | 0.544 | $1.96/Y_{AN}^{Sca}$ | | 1 | | | |
| 3 | | $i_{NBM}-i_{NXI}f_i$ | | | | $1-f_i$ | -1 | | | $f_i$ |
| 4 | | $-i_{NBM}-1/Y_{AN}^{Bro}$ | $-0.87/Y_{AN}^{Bro}-0.76$ | 0.544 | $1.96/Y_{AN}^{Bro}$ | | | 1 | | |
| 5 | | $i_{NBM}-i_{NXI}f_i$ | | | | $1-f_i$ | | -1 | | $f_i$ |
| 6 | $-1/Y_H$ | $-i_{NBM}$ | $-(1-Y_H)/1.71Y_H$ | | $(1-Y_H)/1.71Y_H$ | | | | 1 | |
| 7 | $-1/Y_H$ | $-i_{NBM}$ | | $-(1-Y_H)/2.86Y_H$ | $(1-Y_H)/2.86Y_H$ | | | | 1 | |
| 8 | | $i_{NBM}-i_{NXI}f_i$ | | | | $1-f_i$ | | | -1 | $f_i$ |

Table C3: Biofilm, mass transfer, diffusion, and geometrical parameters values from our previous study

| Coefficient | | Value/range | Unit |
|---|---|---|---|
| $D_{NH4}$ | diffusivity coefficient for ammonium | $6.25 \times 10^{-6}$ | $m^2 h^{-1}$ |
| $D_{NO2}$ | diffusivity coefficient for nitrite | $5.83 \times 10^{-6}$ | $m^2 h^{-1}$ |
| $D_{NO3}$ | diffusivity coefficient for nitrate | $5.83 \times 10^{-6}$ | $m^2 h^{-1}$ |
| $D_{O2}$ | diffusivity coefficient for oxygen | $9.17 \times 10^{-6}$ | $m^2 h^{-1}$ |
| $D_{N2}$ | diffusivity coefficient for dinitrogen | $9.17 \times 10^{-6}$ | $m^2 h^{-1}$ |
| $D_{S\_S}$ | diffusivity coefficient for readily biodegradable substrate | $4.16 \times 10^{-6}$ | $m^2 h^{-1}$ |
| $D_{P,M}$ | Pore and matrix diffusivity coefficient of particulate matters | 0.0001 | $m^2 h^{-1}$ |
| $LL*$ | External mass transfer boundary layer thickness around biofilm | $7.2 \times 10^{-9}$ | m |
| $rho*$ | Biomass density in the biofilm matrix (solid phase) | 148,000 | $g\,COD/m^3$ |
| $r_{sp}{}^*$ | Initial biofilm thickness (initial radius size of one spherical anammox granule) | 0.0005 | m |
| $r_r{}^*$ | Size of a single spherical particle (support cell) | $3.2 \times 10^{-8}$ | m |
| $c*$ | Empirical correction factor | 0.14 | - |

*These values were used for sensitivity analysis and error contribution analysis as well. The range for LL is from 0 to $10^{-4}$ m. The range for rho is from 125,000 to 400,000 g $COD/m^3$. The range for $r_{sp}$ is from 0.00005 to 0.02 and for rr is from $10^{-5}$ to $10^{-8}$.

Table C4: Range of kinetic and stoichiometric parameters for sensitivity analysis and calibration obtained from literatures

| Parameter | Definition | Typical range | Unit |
|---|---|---|---|
| $b_{AN\_Sca,}$ $b_{AN\_Bro}$ | decay rate coefficient of anammox | 0.000045 – 0.000337 | $h^{-1}$ |
| $K_O^{AN}$ | $S_{O2}$ inhibiting coefficient for anammox | 0.01 – 0.4 | $g\ O_2\ m^{-3}$ |
| $Y_H$ | anoxic yield coefficient | 0.43 – 0.67 | $g\ COD\ g^{-1}\ COD$ |
| $\mu_H$ | maximum growth rate | 0.125-0.5 | $h^{-1}$ |
| $b_H$ | decay rate coefficient | 0.008 – 0.05 | $h^{-1}$ |
| $k_H$ | hydrolysis rate constant | 0.05-0.15 | $h^{-1}$ |
| $K_X$ | hydrolysis saturation constant | 0.03-0.1 | $g\ COD\ g^{-1}\ COD$ |
| $K_S$ | $S_S$ affinity constant | 2.0 – 20.0 | $g\ COD\ m^{-3}$ |
| $K_{NOx}^{H}$ | $S_{NOx}$ affinity constant | 0.1-0.5 | $g\ N\ m^{-3}$ |
| $\eta_{NOX}$ | anoxic reduction factor of growth of | 0.16-0.83 | — |
| $i_{NBM}$ | Nitrogen content of biomass | 0.078 – 0.08 | $g\ N\ g^{-1}\ COD$ |
| $i_{NXI}$ | Nitrogen content of inert biomass $X_I$ | 0.02 – 0.06 | $g\ N\ g^{-1}\ COD$ |
| $f_I$ | Fraction of biomass decaying into inert | 0.08-0.1 | $g\ COD\ g^{-1}\ COD$ |

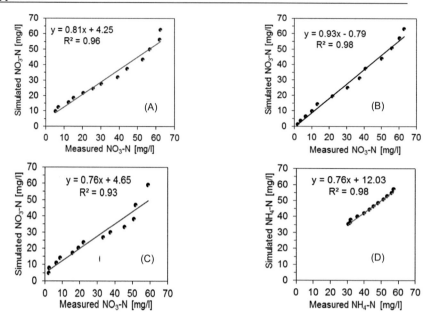

Fig. C2: Analysis of regressions for nitrate nitrogen for assay 2 in the absence of glucose and presence of nitrate (A), assay 3 in the presence of glucose and nitrate (B) and assay 4 in the presence of nitrate and ammonium (C, D)

# Appendix D
## Sensitivity analysis and parameters estimation procedure

To calibrate the models, numerical global sensitivity analysis was performed, in which 65 and 45 simulations was run for the model with and without EPS respectively. The sensitive parameters for the concentrations of $NH_4$, $NO_2$, $NO_3$, COD in the effluent from the bulk phase and the fraction of anammox, AOBs, NOBs, heterotrophs in the biofilm matrix were ranked. The most sensitive parameters for the model with and without EPS and the average values (considering the absolute value and the sign) for the calculated array of sensitivity functions are shown in Fig. 18 for the bulk and Fig. 19 for the solid matrix.

For the bulk phase, parameters including $Y_H$, $Q_{in}$, $\eta_{NOX}$ and $\mu_{Hmax}$ showed highly sensitivity, with the average of sensitivity value varying from 0.1 to 281 g N.m$^{-3}$ or g COD.m$^{-3}$ (Fig. 18) The yield coefficient for heterotrophic bacteria ($Y_H$) negatively influenced on the concentration of $NH_4$ and COD. The sensitivity value of $Y_H$ conducted for the model without EPS was double than model with EPS. But the influence of $Y_H$ on $S_{NO2}$ and $S_{NO3}$ on the model with EPS and without EPS was lower but in the reverse direction than $S_{NH4}$. Referring to the stoichiometric matrix provided in the Table 2A of the appendix it is obviously seen that higher heterotrophic yield coefficient will lead to a lower concentration of the nitrite and nitrate due to the activity of the denitrifying bacteria.

Similar to $Y_H$, anoxic reduction factor ($\eta_{NOX}$) of heterotrophic growth and the maximum growth rate of heterotrophic ($\mu_{Hmax}$) had significant (negative) impact to the concentration of substrates in bulk liquid phase, particularly for $S_{NO2}$, and $S_{NO3}$ (Fig. 18).

The average influent flow rate to the biofilm compartment ($Q_{in}$), which is assumed constant in this study was the third sensitive parameter for all variables of the bulk liquid in both models with and without EPS. This parameter contributed positively on the sensitivity function for $S_{COD}$, $S_{NO2}$, $S_{NO3}$, but a negative value for $S_{NH4}$.

Besides these common sensitive parameters, the nitrogen content of the biomass ($i_{NBM}$) was sensitive for $S_{NH4}$, because during the decay of biomass, soluble organic nitrogen content was converted (mostly) to the form of $NH_4$ and finally contributes to the concentration of $S_{NH4}$ in the effluent. Higher nitrogen content of the biomass will lead to a lower quality of the effluent. Therefore, the calibration of this parameter highly influences on the prediction ability of the mathematical model.

The physical adsorption rate constant for readily biodegradable substrate ($k_{dSs}$) was highly negatively sensitive $S_s$ since the adsorption process of COD on GAC was based on a simplified linear Freundlich equation. It is assumed in the model that physical adsorption occurs simultaneously with other biological substrate removal processes. Thus, adsorption process will inversely proportional to the concentration of $S_{COD}$ in the bulk.

The model considering kinetic formation of EPS and microbial soluble products had two additional sensitive parameters: yield coefficient for growth of heterotrophs on UAP ($Y^H_{UAP}$), and yield coefficient for formation of EPS on heterotrophs ($k^H_{EPS}$). The kinetic of UAP

(167)

formation and uptake during the growth of microorganism had more significant influent on the effluent of wastewater quality compared to BAP which is in compliance with other studies (Barker and Stuckey, 1999).

In addition, $Y^H_{UAP}$ and $Y_H$ were both significant sensitive parameters for the model with EPS. The growth of heterotrophs considers a complicated set of six processes than model without EPS considering two subsequent processes over nitrite and nitrate reduction.

For the biofilm solid matrix, the parameters such as $Y_H$, $\eta_{NOX}$ and $\mu_{Hmax}$ were highly sensitive for the bulk phase but not as sensitive for anammox ($X_{AN}$), AOBs ($X_{AOB}$), NOB ($X_{NOB}$) and heterotrophs ($X_H$), with an average of sensitivity function value varied from 0.01 x $10^3$ to 8.23 x $10^4$ g COD.m$^{-3}$ (Fig. 19). Also, the bacterial density within the biofilm (rho) (including active and inactive organic components) in terms of chemical oxygen demand (COD) positively influences to the concentration of living biomass which average sensitivity value of model with EPS was higher than model without EPS, especially for $X_{AN}$ since anammox was dominant bacterial group within the biofilm.

The density of inorganic carbon compounds within the biofilm matrix formed by adsorption and accumulated over GACs (rho$_{inorganic}$) showed positive impact on the $X_{AOB}$. The rho$_{inorganic}$ was defined to separate the physical GAC adsorption processes from biofilm matrix, thus this parameter requires experimental quantification.

Besides, initial fraction of each bacterial group (fr$_{ini,AN}$, fr$_{ini,AOB}$, fr$_{ini,NOB}$) positively influenced to the corresponding bacterial concentration. It is obvious that initial fraction had significant influence on the results of microbial components, because initial concentration of each group of bacteria was calculated from multiplying of bacterial density (rho) to the initial fraction of corresponding group of bacteria. However, it is notable that although a long-term simulation period was chosen but the sensitivity output concentration of bacteria to the initial fraction is still high.

Another important parameter sensitive for the microbial components was the initial biofilm thickness ($r_{sp}$), because the higher $r_{sp}$ consequences to the higher initial biofilm area and the higher interaction between biofilm and substrate.

The global sensitivity analysis for $S_{NH4}$, $S_{NO2}$, $S_{NO3}$ and $S_{COD}$ in the bulk liquid phase and $X_{AN}$, $X_{AOB}$, $X_{NOB}$ and $X_H$ in the solid matrix was performed in order to provide better insight of which set of parameters influence more to the concentration of substrates and solid particulate matters in the biofilm. These results will be useful for further step of model calibration.

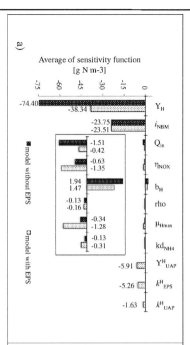

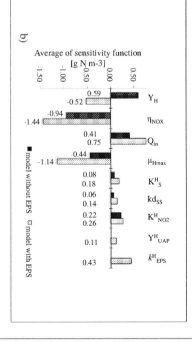

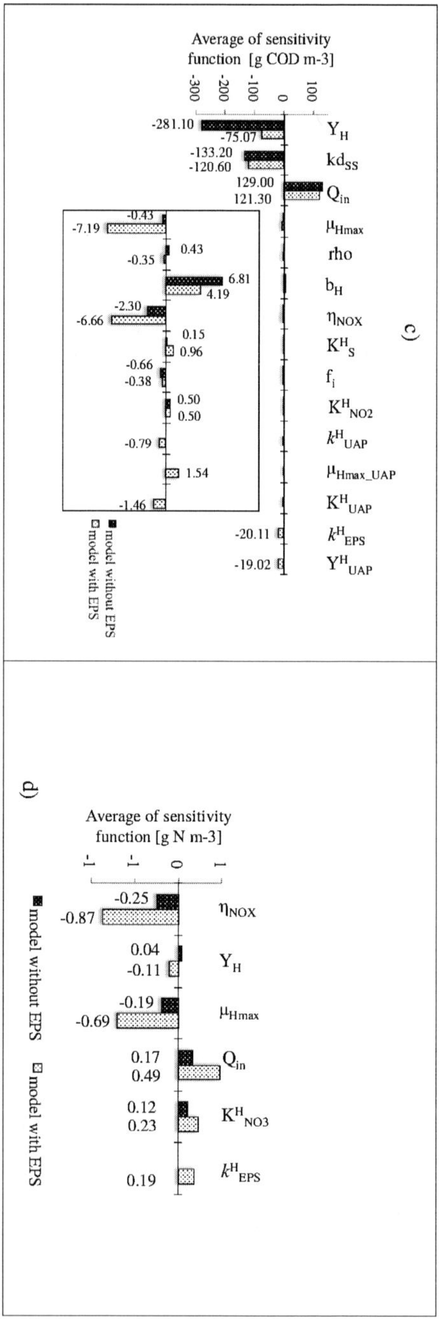

Figure D1: Ranking of sensitive parameters based on calculated average for absolute-relative sensitivity array function (absolute values and signs) for NH4-N (a), NO2-N (b), S-S (c) and S_NO3 (d) for two sort of model with and without EPS

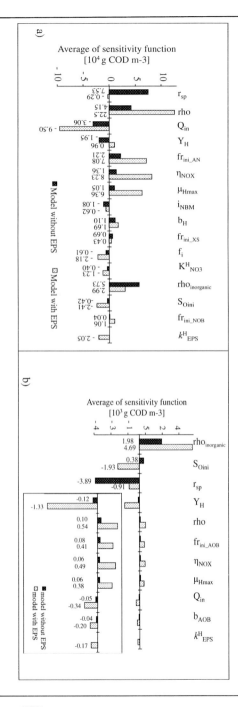

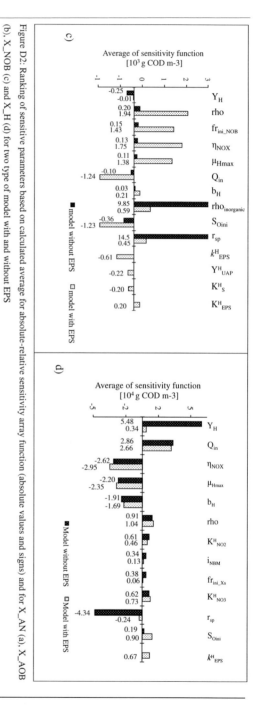

Figure D2: Ranking of sensitive parameters based on calculated average for absolute-relative sensitivity array function (absolute values and signs) and for X_AN (a), X_AOB (b), X_NOB (c) and X_H (d) for two type of model with and without EPS

## Parameter estimation

Results from the sensitivity analysis will be used for parameter estimation to calibrate both models with and without EPS. In this work, we have combined the manual and automated calibration for estimating parameters within the identified ranges based on literatures showed in the Table A3 and A4 of the appendix. During automated calibration, a set of random simulations using Simplex nonlinear parameter estimation algorithm was performed by minimizing the chi-square ($\chi 2$) values between the model data and the actual data of nitrite, nitrate and ammonium-nitrogen. During the calibration procedure, the general prediction trend of the model was visually evaluated and R-squared index ($R^2$) for the regression analysis between simulation and observation was also checked and calculated in a regular basis. The full list of leading to obtain a good efficiency for the calibration period (345 days) with lowest Chi-squared and highest R-squared are presented in Table 8 for the biofilm and mass transfer parameters and in Table 9 for the stoichiometric and kinetic parameters.

With regards to biofilm parameters, the biofilm density (rho) in terms of chemical oxygen demand (COD) was set as 450 000 gCOD m-3 which is agreement with other studies (Ni et al., 2014, 2009). Since the biomass densities for all type of active bacteria and non-active microbial components were set to be the same in our model (rho), hence, the model output of the percentage of bacterial abundance should be equal to their volume fraction of. In this way, the obtained qFISH results based on optimized protocol could be applied to verify the model predictions.

The initial average granule radius ($r_{sp}$) was calibrated as 0.55 mm. The order of magnitude of the biofilm thickness values calculated by the model complies with our previous measurements (Azari et al., 2016) This diffusion bound layer (LL) was calibrated with the thickness at 143 μm complying with other finding which used a calibrated value of 100 μm (Azari et al., 2016; Boltz et al., 2010). With regards to initial parameters, due to reported values, very low oxygen concentration (0.03 - 0.06 mgO$_2$ L$^{-1}$) was considered for the concentration of the oxygen in the bull liquid during the simulation. The initial fractions of bacteria may have an influence on the warm-up period and they were carefully calibrated based within the range identified by previous studies (Azari et al., 2016). Initial input data for the beginning time step were measured according to the collected data for the concentration in the initial tie step. This consists of initial concentration of substrates in aqueous solution (initial concentrations of NH$_4$-N, NO$_2$-N, NO$_3$-N and COD).

## Table D1: Stoichiometric matrix of the model

| Process No | $S_{UAP}$ (COD) | $S_{BAP}$ (COD) | $S_S$ (COD) | $S_O$ (O) | $S_{NH}$ (N) | $S_{N2}$ (N) | $S_{NO3}$ (N) | $S_{NO2}$ (N) | $X_S$ (COD) | $X_{AN}$ (COD) | $X_{AOB}$ (COD) | $X_{NOB}$ (COD) | $X_{EPS}$ (COD) | $X_H$ (COD) | $X_I$ (COD) | $X$ (COD) |
|---|---|---|---|---|---|---|---|---|---|---|---|---|---|---|---|---|
| 1 | | $1-f_S$ | $f_S$ | | | | | | | | | | | | | |
| 2 | | | $1$ | $0$ | $-\frac{1}{Y_{AN}}$ | $\left(-\frac{1}{Y_{AN}}\right)-\left(\frac{1}{1.14}\right)$ | $0.55$ | $\frac{2}{Y_{AN}}$ | | | | | | | $-1$ | |
| 3 | $f_{UAP}^{AN}$ | | | | $-i_{NBM,AN}$ | $\frac{1}{Y_{AN}}$ | | | $1-f_I-f_{BAP}$ | $1-f_{UAP}^{AN}-f_{EPS}^{AN}$ | | | $f_{EPS}^{AN}$ | | $\frac{k_{EPS}^{AN}}{...}$ | $f_I$ |
| 4 | $f_{UAP}^{AOB}$ | | | $-\frac{3.43-Y_{AOB}}{Y_{AOB}}$ | $-i_{NBM}-\frac{1}{Y_{AOB}}$ | | | $\frac{1}{Y_{AOB}}$ | | | $1-f_{UAP}^{AOB}-f_{EPS}^{AOB}$ | | $f_{EPS}^{AOB}$ | | | $f_{UAP}^{AN}$ |
| 5 | $f_{UAP}^{NOB}$ | | | $-\frac{1.14-Y_{NOB}}{Y_{NOB}}$ | $-i_{NBM}$ | $-\frac{1}{Y_{NOB}}$ | $\frac{1}{Y_{NOB}}$ | | | | | $1-f_{UAP}^{NOB}-f_{EPS}^{NOB}$ | $f_{EPS}^{NOB}$ | | | |
| 6 | | $f_{BAP}$ | | | $i_{NBM}-(f_I\times i_{NXI})$ | | | | $1-f_I-f_{BAP}$ | $-i_{NBM}$ | | | | | $\frac{k_{EPS}^H}{Y_H}$ | $f_I$ |
| 7 | | $f_{BAP}$ | | | $-(f_I\times i_{NXI})$ | | | | $1-f_I-f_{BAP}$ | | $-i_{NBM}$ | | | | $\frac{k_{EPS}^H}{Y_H}$ | $f_I$ |
| 8 | $f_{BAP}$ | | | | $-(f_I\times i_{NXI})$ | $-\frac{1}{Y_{NOB}}$ | $\frac{1}{Y_{NOB}}$ | | $1-f_I-f_{BAP}$ | | | $-i_{NBM}$ | | | $\frac{k_{EPS}^H}{Y_H}$ | $f_I$ |
| 9 | | $f_{BAP}$ | | | $-i_{NBM}$ | $-\frac{(1-Y_H)(1-k_{UAP}^H-k_{EPS}^H)}{1.71\times Y_H}$ | $-\frac{(1-Y_H)(1-k_{UAP}^H-k_{EPS}^H)}{2.86\times Y_H}$ | $-\frac{(1-Y_H)(1-k_{UAP}^H-k_{EPS}^H)}{1.71\times Y_H}$ | | | | | | $1-k_{UAP}^H-k_{EPS}^H$ | $\frac{k_{EPS}^H}{Y_H}$ | $f_I$ |
| 10 | $\frac{k_{UAP}^H}{Y_H}$ | | $-\frac{1}{Y_H}$ | $-\frac{1-Y_H}{Y_H}$ | $-i_{NBM}$ | $-\frac{1-Y_H}{2.86\times Y_H}$ | $-\frac{1-Y_H}{2.86\times Y_H}$ | $-\frac{1-Y_H}{1.71\times Y_H}$ | | | | | | $1$ | $\frac{k_{EPS}^H}{Y_H}$ | $f_I$ |
| 11 | $\frac{k_{UAP}^H}{Y_H}$ | $-1$ | | | $-i_{NBM}$ | $-\frac{1-Y_{BAP}^H}{2.86\times Y_{BAP}^H}$ | $-\frac{1-Y_{BAP}^H}{2.86\times Y_{BAP}^H}$ | $-\frac{1-Y_{BAP}^H}{1.71\times Y_{BAP}^H}$ | | | | | | | $\frac{k_{EPS}^H}{Y_H}$ | $1$ |
| 12 | | $-\frac{1}{Y_{BAP}^H}$ | | | $-i_{NBM}$ | $-\frac{1-Y_{BAP}^H}{2.86\times Y_{BAP}^H}$ | $-\frac{1-Y_{BAP}^H}{2.86\times Y_{BAP}^H}$ | $-\frac{1-Y_{BAP}^H}{1.71\times Y_{BAP}^H}$ | | | | | | | $1$ | $1$ |
| 13 | $-\frac{1}{Y_{UAP}^H}$ | | | | $-i_{NBM}$ | $-\frac{1-Y_{UAP}^H}{1.71\times Y_{UAP}^H}$ | $-\frac{1-Y_{UAP}^H}{2.86\times Y_{UAP}^H}$ | | | | | | | $1$ | $1$ | $1$ |
| 14 | $-\frac{1}{Y_{UAP}^H}$ | $-\frac{1}{Y_{BAP}^H}$ | | | $-i_{NBM}$ | $-\frac{1-Y_{UAP}^H}{1.71\times Y_{UAP}^H}$ | $-\frac{1-Y_{UAP}^H}{2.86\times Y_{UAP}^H}$ | $-\frac{1-Y_{UAP}^H}{2.86\times Y_{UAP}^H}$ | $-i_{NBM}$ | | | | | | $-1$ | $-1$ |
| 15 | | $f_{BAP}$ | | | $\begin{array}{c}i_{NBM}\\-(f_I\\\times i_{NXI})\end{array}$ | | | | | | | | | $-1$ | | $f_I$ |

Table D2: Biofilm compartment characteristics

| Parameter | Description | Value/range | Unit | Source |
|---|---|---|---|---|
| $D_{BAP}^*$ | Diffusivity coefficient for biomass-associated products (BAP) | $1.38 \times 10^{-5}$ | $m^2 d^{-1}$ | Liu et al. 2016 |
| $D_{UAP}^*$ | Diffusivity coefficient for utilization-associated products (UAP) | $1.38 \times 10^{-5}$ | $m^2 d^{-1}$ | Liu et al. 2016 |
| $D_{NH4}$ | Diffusivity coefficient for ammonium | $1.50 \times 10^{-5}$ | $m^2 d^{-1}$ | Williamson & McCarty 1976 |
| $D_{NO2}$ | Diffusivity coefficient for nitrite | $1.40 \times 10^{-5}$ | $m^2 d^{-1}$ | Williamson & McCarty 1976 |
| $D_{NO3}$ | Diffusivity coefficient for nitrate | $5.83 \times 10^{-6}$ | $m^2 h^{-1}$ | Williamson & McCarty 1976 |
| $D_{O2}$ | Diffusivity coefficient for oxygen | $2.20 \times 10^{-5}$ | $m^2 d^{-1}$ | Picioreanu et al. 1997 |
| $D_{N2}$ | Diffusivity coefficient for dinitrogen | $2.20 \times 10^{-5}$ | $m^2 d^{-1}$ | Williamson & McCarty 1976 |
| $D_{S\_S}$ | Diffusivity coefficient for readily biodegradable substrate | $1.38 \times 10^{-5}$ | $m^2 d^{-1}$ | Winkler et al. 2015 |
| LL | External mass transfer boundary layer thickness | $0–10^{-4}$ | m | Boltz et al. 2010 |
| rho | Biomass density in the biofilm matrix (solid phase) | 20,000–450,000 | g COD $m^3$ | Boltz et al. 2010, Ni et al 2009 |
| $r_{sp}$ | Initial biofilm thickness (average radius size of one spherical anammox granule) | 0.00025–0.01 | m | Azari et al. 2017 Ke et al. 2015 |
| $r_R$ | Size of support particle (one cell) | $10^{-8}–10^{-5}$ | m | Reichert 1998, Winkler et al. 2015 |
| $Q_{in}$ | Range of daily influent flow rate to biofilm compartment | 350–570 | $m^3.d^{-1}$ | Data collected in this study |
| $V_{tot}$ | Total working volume of biofilm compartment | 130–180 | $m^3$ | Data collected in this study |

* Parameters are only available in the model with EPS.

Table D3: Calibrated biofilm and mass transfer parameters

| Parameter | Description | Value | Unit |
|---|---|---|---|
| c | Correction factor for effective biofilm area | 0.27 | - |
| LL | External mass transfer boundary layer thickness | $1.43 \times 10^{-6}$ | m |
| $n_{sp}$ | Number of spherical particles | 415380 | - |
| rho | Biomass density in the biofilm matrix (solid phase) | 450000 | g COD $m^3$ |
| rho_adsorbed | Density of inorganic adsorbant in form of carbon | 450000 | g COD $m^3$ |
| rho_inorganic_N | Density of inorganic adsorbant in form of nitrogen | 5000 | g N $m^3$ |
| $r_{sp}$ | Initial biofilm thickness (average radius size of one spherical anammox granule) | 0.00055 | m |
| $r_R$ | Size of the support particle (one cell) | $4.78 \times 10^{-8}$ | m |
| $Q_{in}$ | Average influent flow rate to biofilm compartment | 350 | $m^3.d^{-1}$ |

Table D4: Calibrated kinetic and stoichiometric parameters

| Parameter | Definition | Set value | Unit |
|---|---|---|---|
| **Kinetic parameters for anaerobic ammonium-oxidizing bacteria (X_AN)** | | | |
| $b_{AN}$ | Anoxic decay rate coefficient of X_AN | 0.0024 | $d^{-1}$ |
| $K_{NH_4}^{AN}$ | $S_{NH4}$ affinity constant for X_AN | 0.2 | g N m$^{-3}$ |
| $K_{NO_2}^{AN}$ | $S_{NO2}$ affinity constant for X_AN | 0.3 | g N m$^{-3}$ |
| $K_{O_2}^{AN}$ | $S_{O2}$ inhibiting coefficient for X_AN | 0.11 | g O m$^{-3}$ |
| $\mu_{ANmax}$ | Maximum growth rate of X_AN | 0.23 | $d^{-1}$ |
| $fr_{ini\_AN}$ | Initial fraction of X_AN in biofilm solid matrix | 0.25 | - |
| **Kinetic parameters for ammonium-oxidizing bacteria (X_AOB)** | | | |
| $b_{AOB}$ | Anoxic decay rate coefficient of X_AOB | 0.065 | $d^{-1}$ |
| $\mu_{AOBmax}$ | Maximum growth rate of X_AOB | 1.9 | $d^{-1}$ |
| $K_{NH_4}^{AOB}$ | $S_{NH4}$ affinity constant for X_AOB | 0.22 | g N m$^{-3}$ |
| $K_{O_2}^{AOB}$ | $S_{O2}$ inhibiting coefficient for X_AOB | 0.76 | g O m$^{-3}$ |
| $fr_{ini\_AOB}$ | Initial fraction of X_AOB in biofilm solid matrix | 0.03 | - |
| **Kinetic and initial parameters for nitrite-oxidizing bacteria (X_NOB)** | | | |
| $b_{NOB}$ | Anoxic decay rate coefficient of X_NOB | 0.002 | $d^{-1}$ |
| $\mu_{NOBmax}$ | Maximum growth rate of X_NOB | 1.8 | $d^{-1}$ |
| $K_{NH_4}^{NOB}$ | $S_{NH4}$ affinity constant for X_NOB | 0.04 | g N m$^{-3}$ |
| $K_{NO_2}^{NOB}$ | $S_{NO2}$ affinity constant for X_NOB | 1 | g N m$^{-3}$ |
| $K_{O_2}^{NOB}$ | $S_{O2}$ inhibiting coefficient for X_NOB | 1.5 | g O m$^{-3}$ |
| $fr_{ini\_NOB}$ | Initial fraction of X_NOB in biofilm solid matrix | 0.04 | - |
| **Kinetic parameters for heterotrophic bacteria (X_H)** | | | |
| $b_H$ | Decay rate coefficient of X_H | 0.251 | $d^{-1}$ |
| $\mu_{Hmax}$ | Maximum growth rate of X_H | 7.5 | $d^{-1}$ |
| $\mu_{Hmax\_BAP}$ | Maximum growth rate of X_H on BAP | 0.072 | $d^{-1}$ |
| $\mu_{Hmax\_UAP}$ | Maximum growth rate of X_H on UAP | 6 | $d^{-1}$ |
| $\eta_{NOX}$ | Anoxic reduction factor of growth of X_H | 0.85 | - |
| $K_{BAP}^{H}$ | Half saturation constant for heterotroph growth on BAP | 85 | g COD.m$^{-3}$ |
| $K_H$ | Hydrolysis rate constant | 3.1 | $d^{-1}$ |
| $K_X$ | Hydrolysis saturation constant | 0.03 | g COD.g$^{-1}$COD |
| $K_S^{H}$ | $S_S$ affinity constant for X_H | 10 | g COD m$^{-3}$ |
| $K_{NO_2}^{H}$ | $S_{NO2}$ affinity constant for X_H | 0.3 | g N m$^{-3}$ |
| $K_{NO_3}^{H}$ | $S_{NO3}$ affinity constant for X_H | 0.3 | g N m$^{-3}$ |
| $K_{NH_4}^{H}$ | $S_{NH4}$ affinity constant for X_H | 0.009 | g N m$^{-3}$ |
| $K_{O_2}^{H}$ | $S_{O2}$ inhibiting coefficient for X_H | 0.2 | g O m$^{-3}$ |
| $K_{UAP}^{H}$ | $S_{UAP}$ affinity constant for X_H | 50 | g COD m$^{-3}$ |
| $fr_{ini\_H}$ | Initial fraction of X_H in biofilm solid matrix | 0.25 | - |

Kinetic parameters for microbial products (X_EPS)

| | | | |
|---|---|---|---|
| $fr_{ini\_EPS}$ | Initial fraction of X_EPS in biofilm solid matrix | 0.2 | - |
| $K_{H\_EPS}$ | Hydrolysis rate constant for EPS | 0.18 | $d^{-1}$ |

Stoichiometric and physiological parameters

| | | | |
|---|---|---|---|
| $i_{NBM}$ | Nitrogen content in total active biomass in granular activated sludge including all heterotrophs and autotrophs | 0.08 | g N $g^{-1}$ COD |
| $i_{NBM\_AN}$ | Nitrogen content in anammox in granular activated sludge | 0.07 | g N $g^{-1}$ COD |
| $i_{NXI}$ | Nitrogen content of inert biomass $X_I$ | 0.02 – 0.07 | g N $g^{-1}$ COD |
| $f_i$ | Fraction of biomass decaying into inert | 0.08 | g COD $g^{-1}$ COD |
| $f_{BAP}$ | Fraction of BAP decay in biomass | 0.0215 | - |
| $f_S$ | Fraction of EPS converted to readily biodegradable substrates | 0.08 | - |
| $f_{EPS}^{AN}$ | Fraction of EPS produced for anammox | 0.08 | g COD.$g^{-1}$N |
| $f_{EPS}^{AOB}$ | Fraction of EPS produced for AOB | 0.15 | g COD.$g^{-1}$N |
| $f_{EPS}^{NOB}$ | Fraction of EPS produced for NOB | 0.12 | g COD.$g^{-1}$N |
| $f_{UAP}^{AN}$ | Fraction of UAP produced for anammox | 0.05 | g COD $g^{-1}$ N |
| $f_{UAP}^{AOB}$ | Fraction of UAP produced for AOB | 0.17 | g COD $g^{-1}$ N |
| $f_{UAP}^{NOB}$ | Fraction of UAP produced for NOB | 0.16 | g COD $g^{-1}$ N |
| $k_{UAP}^{H}$ | Yield coefficient for UAP of X_H | 0.06 | - |
| $k_{EPS}^{H}$ | Yield coefficient for EPS of heterotroph | 0.2 | - |
| $Y_H$ | Yield coefficient for X_H growth on S_S | 0.67 | g COD $g^{-1}$ COD |
| $Y_{BAP}^{H}$ | Yield coefficient for X_H growth on S_BAP | 0.5 | g COD $g^{-1}$ COD |
| $Y_{UAP}^{H}$ | Yield coefficient for X_H growth on S_UAP | 0.6 | g COD $g^{-1}$ COD |
| $Y_{AN}$ | Yield coefficient for X_AN | 0.16 | g COD $g^{-1}$ N |
| $Y_{AOB}$ | Yield coefficient for X_AOB | 0.19 | g COD $g^{-1}$ N |
| $Y_{NOB}$ | Yield coefficient for X_NOB | 0.05 | g COD.$g^{-1}$ N |

Physical adsoption kinetic parameters

| | | | |
|---|---|---|---|
| $k_d^{NH4}$ | Adsorption factor of ammonium | 0.002 | - |
| $k_d^{SS}$ | Adsorption factor of COD | 8.6 | - |

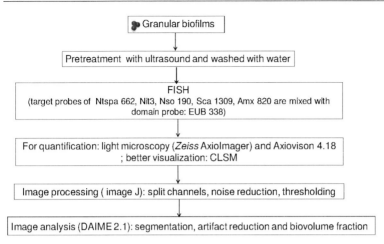

Figure D7: The scheme of qFISH experiments to measure the relative abundance of bacterial groups.

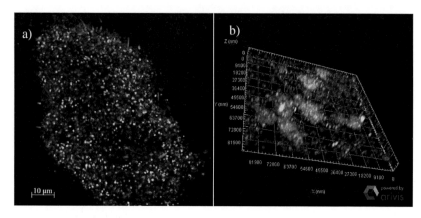

Figure D3: 2D FISH micrographs (a) and 3D FISH micrograph (b) with: AMX820 specific probe in Cy3 dye (orange) for granules from biofilm reactor hybridized with 10 % formamide to target species Ca. Brocadia anammoxidans and Ca. Kuenenia stuttgartiensis. EUB 338 probe in FITC dye (green) was used to target most domain bacteria. Combined signals are in yellow. Imaging done with a Zeiss CLSM 510 for the sample taken on 2 September (c).

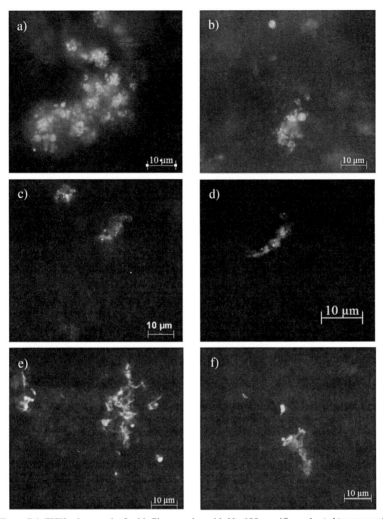

Figure D4: FISH micrographs for biofilm samples with Nso190 specific probe (a,b) to target βAOB, Ntspa662 specific probe (c,d) to target Nitrospira genus of NOBs and NIT3 specific probe (e,f) to target Nitrobacter genus of NOBs. Specific probes are in Cy3 (orange) and EUB 338 probe in FITC dye (green) was used to target most of domain bacteria. Combined signals are in yellow. Imaging done using epifluorescence microscopy for samples taken on 28th July (a,c,e) and samples taken on 2nd September (b,d,f). Magnification ×100 for all images.

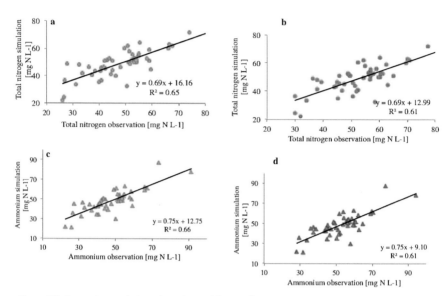

Figure D5: Regression analysis and goodness of fitness $R^2$ for weekly simulated total nitrogen effluent versus observation of model with (a) and without EPS (b), and weekly simulated ammonium effluent versus observation of model with (c) and without EPS (d).

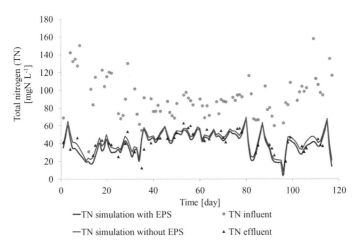

Figure D6. Time-dependent graph for total nitrogen concentration during the validation period for the observed influent, observed effluent and the simulated effluent corresponding to two types of models with and without EPS.

## Appendix E

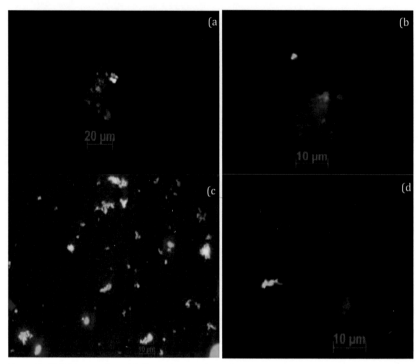

Figure E: FISH images from (a) K1 carriers on day 187 with Ntspa662 target probe (b) K1 carriers on day 187 with Nso190 target probe (c) Kamen's sludge SBR 1 on day 157 with Nso190 target probe (d) from ZDA's sludge on day 27 with Amx820 target probe.

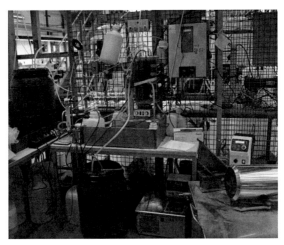

Figure E.2. Lab-scale IFAS SBR on day 155.

Figure 3. Plant equipments (a) IFAS-SBR on day 155 (b) thermostat for temperature reduction.